Organizing
for
Mathematics
Instruction

Organizing for Mathematics Instruction

1977 Yearbook

F. Joe Crosswhite
1977 Yearbook Editor
Ohio State University

Robert E. Reys
General Yearbook Editor
University of Missouri

National Council of
Teachers of Mathematics

Library of Congress Cataloging in Publication Data:

Main entry under title:

Organizing for mathematics instruction.

(Yearbook—National Council of Teachers of Mathematics; 1977)
Includes bibliographies.
1. Mathematics—Study and teaching—Addresses, essays, lectures. I. Crosswhite, F. Joe. II. Series: National Council of Teachers of Mathematics. Yearbook; 1977.
QA1.N3   1977   [QA11]   510′.7s   [510′.7′1]   77-23294

Table of Contents

   Thomas A. Romberg, University of Wisconsin, Madison,
   Wisconsin

   The Individually Guided Education model involves
   groups of teachers working together as instructional decision-making units rather than individual teachers working
   independently in self-contained classrooms. The organizational target in the model described in this essay is an
   entire elementary school or total elementary system.

   Peggy A. House, University of Minnesota, Minneapolis,
   Minnesota

   A model is described for organizing a secondary school
   mathematics department so that learning activities are
   centralized in a large mathematics resource area. Flexible scheduling, mastery learning, and unit-based curricula are features of this model.

   Janet Barnard, Colene Hoose Elementary School, Normal, Illinois
   Carol Dodd Thornton, Illinois State University, Normal, Illinois

   Ways are discussed in which teachers can individualize
   inexpensively within their own classrooms. Examples are
   drawn from the intermediate grades, although the model
   is applicable at any level, grade 4 through high school.

Joseph Abruscato, University of Vermont, Burlington, Vermont
Clinton A. Erb, University of Vermont, Burlington, Vermont

A variety of ways are presented in which independent learning units can be organized and used to supplement regular class instruction.

Robert F. Nicely, Jr., Pennsylvania State University, University Park, Pennsylvania

Based on concepts of diagnostic-prescriptive teaching and mastery learning, a plan is outlined for developing and implementing instructional units. Patterns of achievement on specified sets of objectives are used to define instructional subgroups; the units are intended to provide differential experiences for the smaller groups.

Herbert Fremont, Queens College, Flushing, New York

Maximizing individual progress in mathematics through cooperative effort within subgroups is the subject of this essay. A detailed plan is offered for organizing the class as a cooperative community based on survival groups.

John C. Peterson, Ohio State University, Columbus, Ohio

Simulations are defined as realistic games or models that provide participants with lifelike problem-solving experiences. They also frequently involve the aspects of small-group instruction, although the focus here is more on the organization of the simulation than on the grouping of participants.

Judith Harle Hector, Walters State Community College, Morristown, Tennessee

Organizing to implement a mastery-learning model in a group-based instructional setting is the subject. The model was applied both in a developmental mathematics course at a community college and in an intermediate algebra course at a large state university.

Thomas A. Cooney, University of Georgia, Athens, Georgia

The organization of lessons based on logical considerations arising from the structure of mathematics is discussed. The essay deals with the content of instruction, but its implications have impact for teaching methods as well as curriculum.

Jane Donnelly Gawronski, San Diego County Schools, San Diego, California

Joan E. Fehlen, Minneapolis Public Schools, Minneapolis, Minnesota

> Optional or alternative educational models provide a choice for students, parents, and teachers. Organizational considerations that arise in developing an integrated curriculum within an open school option are discussed.

Gerald R. Rising, State University of New York at Buffalo, Buffalo, New York

Stephen I. Brown, State University of New York at Buffalo, Buffalo, New York

Lawrence N. Meyerson, State University of New York at Buffalo, Buffalo, New York

> The teacher-centered classroom should not be rejected as an alternative. The essay provides a positive view of this type of alternative as it can occur in the hands of a master teacher, and illustrative vignettes from several grade levels are drawn.

Harold L. Schoen, University of Iowa, Iowa City, Iowa

> The research on self-paced instruction is reviewed and implications for instruction drawn. The essay is a potential source for guidance on adopting or implementing such a program.

Max Bell, University of Chicago, Chicago, Illinois

Edward Esty, National Institute of Education, Washington, D.C.

Joseph N. Payne, University of Michigan, Ann Arbor, Michigan

Marilyn N. Suydam, Ohio State University, Columbus, Ohio

> Four reports, which include recommendations on the use of hand-held calculators in schools, are described. Excerpts are drawn from the NACOME report, the Euclid Conference on Basic Skills, a report to the National Science Foundation, and a conference cosponsored by NSF and the National Institute of Education. In a final section, specific activities involving the use of calculators are cited.

Preface

The 1977 Yearbook is the second to be developed under a new concept recommended by the Publications Committee of the National Council of Teachers of Mathematics. The guidelines for these yearbooks specify that each be developed around a central theme that focuses on recent developments, current issues, and future projections, especially as they relate to classroom practices; that the central theme be developed through essays whose treatment may be neither definitive nor exhaustive; and that non-thematic essays whose timely ideas warrant widespread distribution to the Council's members be included. In addition, unlike former yearbooks and the currently developing professional reference series, essays are not commissioned in advance to fulfill a fixed outline determined by an editorial panel; rather, open competition in response to a call for proposals that identifies and delimits the central theme is encouraged. Given these guidelines and the working title Alternatives in Organizing for Mathematics Instruction, we accepted the challenge of developing this yearbook.

Many variables may be manipulated in organizing for instruction—time, space, people, curriculum, instructional materials, teaching methods, and so on. Clearly not all could be treated adequately in a single yearbook. To provide a basis for sharpening the focus in our call for proposals, then, we invited a large number of mathematics educators to respond to the open question, What would you expect to see in an NCTM yearbook entitled Alternatives in Organizing for Mathematics Instruction? Almost fifty people responded. Although their recommendations were wide-ranging, they were very helpful in developing the call for proposals. These persons are too numerous to list individually, but we wish to acknowledge their contribution here.

To delimit the scope of the yearbook and to avoid duplication with past or anticipated yearbooks, we asked that the organization of mathematical content (curriculum) not be the major theme of any essay. Instead, we suggested that a specific teaching approach be selected and that this approach be developed in relation to one or more patterns of administrative organization. Priority was given to the teaching approach because it is the organizing act most directly under the control of the individual teacher. How the approach relates to administrative structures was suggested because such structures represent a real constraint on the teaching act. We further suggested that the essay include examples illustrating the teaching approach in specific mathematical contexts and that theoretical or research rationales be subordinate to practical considerations. Forty outlines were submitted in response to the call for proposals, and on the basis of review, authors were invited to prepare the thematic essays that appear here.

We have elaborated on the process by which the yearbook was generated because we feel it is important that the reader understand the limitations inherent in that process. The yearbook does not constitute a definitive treatment of any single organizational alternative, nor is it comprehensive with respect to the range of alternatives possible. The collection of thematic essays preserves any imbalance that may have existed in the set of proposals submitted. There are areas that might appropriately have been treated but for which no proposals were received. The distribution of proposals over organizational areas perhaps inaccurately reflected the general pattern of concern. In a few areas, proposals were received and essays were invited but do not appear in the final collection for one of several reasons. Finally, we took quite seriously what we understood to be the motivation for choosing the word *essay* to describe the contributions to the new series of yearbooks. An essay has been defined as a short "analytic or interpretive literary composition usually dealing with its subject from a limited or personal point of view." The thematic essays in this yearbook are indeed written from a personal point of view and do treat their subject in a limited way. The authors provide an analysis and interpretation of an organizational alternative that they advocate, but they have been limited, at our direction, in their opportunity to present a research or theoretical base to support their position.

The reader should respect these limitations in interpreting the thematic essays. Each advocates a position that in the judgment and experience of the author represents a viable and effective alternative in organizing for instruction in mathematics. They do not, in any case, necessarily represent a position recommended or endorsed by the National Council of Teachers of Mathematics. The intent of the yearbook is to provide a forum for discussion. Consistent with this intent, we have not restricted the choice of alternatives. The reader should not be surprised to find mutually contradictory positions expressed in two separate essays.

The first eleven essays in the book develop the central theme. In the absence of a preconceived outline, an artificial structure has been imposed to sequence the essays. In a very loose sense, the flow is from the individual student to the full class as the focal point for instruction.

We argued at some length about the classification of the twelfth essay, vacillating between thematic and nonthematic. The essay has thematic characteristics in that it does concern an organizational alternative, although no position is taken. We feel it important to include this essay because it suggests the reasonable idea that caution should be used in considering *any* instructional alternative and that any relevant data accumulated on the subject—even though it may be inadequate—should be consulted prior to adoption or implementation.

The final essay is more clearly nonthematic. We would be hard pressed to identify a topic more nearly meeting the criterion of timeliness than the potential impact of the hand-held calculator on mathematics education.

The successful completion of this yearbook is a direct result of the dedication, talents, enthusiasm, and help of many individuals. In addition to the authors of the essays, many others contributed much time and effort. In particular, we thank those who refereed the manuscripts initially proposed as well as those who reviewed the essays at different stages of development. They provided many valuable suggestions and recommendations that either directly or indirectly improved the quality of the yearbook. We are most appreciative of this help.

Without a permanent editorial panel, we had to rely heavily on feedback from the referees and reviewers for the screening and reviewing of proposed essays. However, we also had periodic review meetings and would be remiss in our acknowledgments if we failed to note the individual contributions made by Jon L. Higgins and Marilyn N. Suydam. They made substantial contributions through all stages of development and refinement, and our deepest thanks go to them for their help. Finally, our sincere appreciation is given to Charles R. Hucka and the production staff of the NCTM for their help in making this yearbook a reality.

It is our hope that the instructional alternatives presented will indeed provide some viable options for different readers. The ultimate reward to all who have contributed to this yearbook will be the extent to which the ideas contained herein result in improving the teaching and learning of mathematics.

F. JOE CROSSWHITE
1977 Yearbook Editor

ROBERT E. REYS
General Yearbook Editor

1

Organizing for Individualization: The IGE Model

Thomas A. Romberg

*O*ne increasingly popular alternative method of organizing instruction in elementary schools is known as Individually Guided Education (IGE). How mathematics instruction is carried out in an IGE setting can be illustrated by developing two basic points: first, *effective* mathematics instruction takes into account the individual needs of students; second, *efficient* mathematics instruction can be accomplished through the Instructional Programming Model developed for IGE.

Effective Instruction

Mathematics is abstract and lifeless without a student. Yet, although students bring life to mathematics, they add to the instructional complexity, for they also bring to the activities the full range of their differences. One cannot assume that the same activities will be equally effective with all children nor even that all children need to learn the same concepts and

This essay reflects research and development conducted by the Wisconsin Research and Development Center for Cognitive Learning, supported in part by funds from the Office of Education, the National Institute for Education, the National Science Foundation, and the Sears Roebuck Foundation.

skills. In fact, if we consider a number of points well supported by existing work in psychology and education, we are forced to conclude that instruction must be adapted for the individual student. And in order to work effectively with each student, we must take note of the following two generalizations.

1. *Students acquire mathematical processes, concepts, and skills in a developmental sequence.*

The work of developmental psychologists suggests that students grow and develop through stages, that this development is cognitive and affective as well as physical, and that physical maturation greatly influences cognitive and affective growth. According to Lovell (1971, pp. 12–13), the bulk of the developmental literature suggests the following educational implications:

a. Pupils should work in small groups or individually at tasks that have been provided.

b. Opportunity should be provided for pupils to act on physical materials or to use games.

c. Social intercourse using verbal language should be encouraged, since it is an important influence in the development of concrete operational thought.

d. The direction of the work is the teachers' responsibility.

e. Alongside the abstraction of the mathematical idea from the physical situation, there must be the introduction of the relevant symbolization and the working of examples, involving drill and practice and problems.

2. *Students achieve at different rates.*

Some students learn almost spontaneously, whereas others learn only after much repetition and effort. The effective way to deal with different achievement rates is to pace instruction accordingly.

These psychological facts indicate only two ways in which students differ from one another. If other psychological, biological, and environmental differences are considered, the school and the teachers are faced with the overwhelming task of providing differential instruction for each student. What is needed, then, is an organized, efficient approach to instruction.

Efficient Instruction

One way to provide efficient mathematics instruction is to follow the principles of Individually Guided Education (IGE). These principles have been described by Klausmeier and others (Klausmeier, Rossmiller, and Saily 1976; Klausmeier 1972).

> IGE is a comprehensive system. It focuses on the individual child as the unit of instruction, rather than the classroom group. Groups of teachers working together form the instructional decision-making units rather than individual teachers working independently in self-contained classrooms. The building staff, including the principal, are the primary educational decision makers for the children attending the particular school rather than the central office staff or the building principal independent of the teachers. Thus the IGE system requires changes in all the aspects of the total educational system. . . . IGE differs from the others in that it is a total system. [Klausmeier 1972, pp. 4–8]

Two components of IGE, the multiunit organization and the instructional programming model, have been developed to facilitate instruction.

The multiunit organization

Both a new instructional pattern and a new administrative pattern at the building and central-office levels are proposed. Figure 1.1 shows the organization for an elementary school having 400 to 600 students. Instruc-

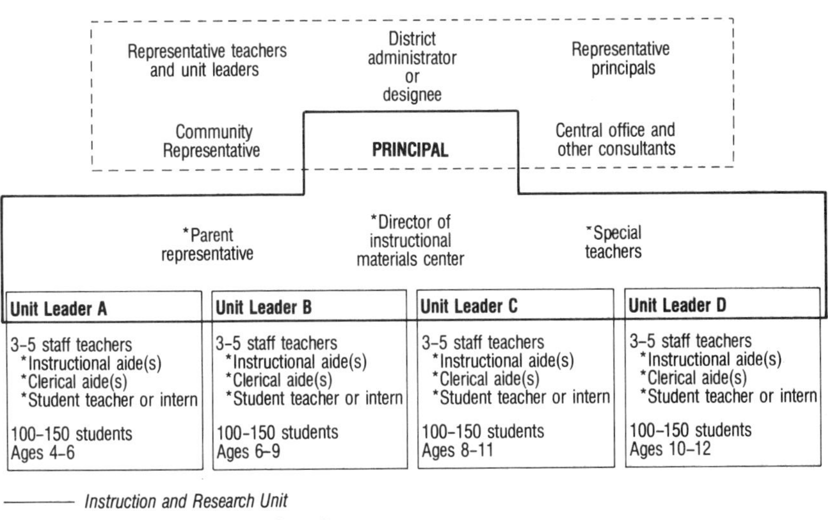

Fig. 1.1. Multiunit organization of an IGE school (adapted from Klausmeier, Morrow, and Walter [1968])

tional and research units (I & R units) replace self-contained classrooms as the instructional organization. A unit leader; three to five teachers; and one or more instructional aides, clerical aides, or student teachers or interns comprise the staff of an ungraded, multiage group of 100 to 150 students. The unit leaders and the building principal form the Instructional Improvement Committee (IIC) of the building and share in making educational decisions. A district administrator, building principals, unit leaders and selected teachers, a community representative, and other central-office personnel make up the System-wide Program Committee (SPC). These new organizational and administrative arrangements are designed to provide for making appropriate educational and instructional decisions; to encourage open communication among students, teachers, and principals; and to permit accountability by educational personnel at the different levels.

The Instructional Programming Model

IIC's and the I & R units use the Instructional Programming Model (IPM) to provide for differences among students in their rates and styles of learning, levels of motivation, and other characteristics (see fig. 1.2).

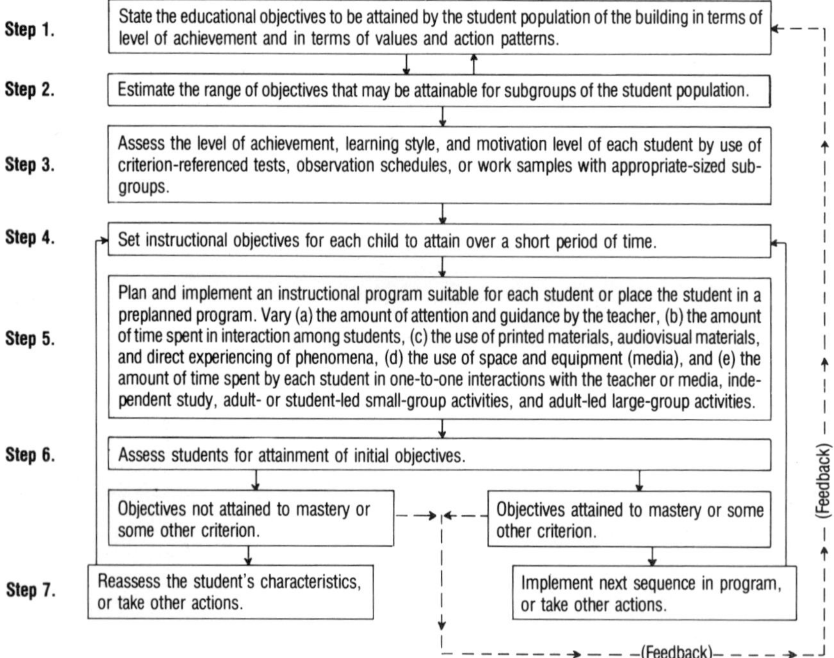

Fig. 1.2. Instructional Programming Model (adapted from Klausmeier et al. [1971, p. 19])

At the center of the IPM is step 5—implementing a plan of instruction. The principles of instruction that underlie this step stem from what we know about individual differences. In practical terms, providing for individual differences amounts to *differentiating instruction*. This involves forming and reforming groups of students to work on different mathematical topics or to work on the same topic in different ways. Only by differentiating both across and within topics will instruction be efficient. To accomplish this, three principles of instruction are assumed: (1) Planning and decision making should be made at several levels by appropriate groups. Steps 1, 2, 4, 5, and 7 of the IPM involve making decisions about instruction for children. Step 1 is at the IIC level, the rest at the I & R unit level. A number of individuals participate at each level, that is, instruction is *guided* by professional teams. (2) Any instructional program should be objective referenced. Note that the word *objective* is explicitly stated for steps 1, 2, 4, and 6 of the IPM. Identifying objectives in mathematics reinforces the teacher's role as a facilitator of instruction. When teachers have stated objectives for mathematics instruction, they have identified what the children are to learn. Thus the objectives provide a starting point for facilitating children's learning. (3) Systematic criterion-referenced assessment is essential (steps 3 and 6 of the IPM). The purpose of assessment is to gather information about a child as an individual. Assessment in IGE is done before the child participates in an instructional sequence, during the sequence, and at the end of the sequence.

The IGE Instructional Programming Model provides a framework on which an *efficient* program of mathematics instruction can be built.

Providing Instruction

Teaching mathematics in an individually guided setting is not simple. Many different tasks must be performed, and many different skills are needed.

Step 1: Identifying objectives

The responsibility for identifying objectives rests with the IIC and is based on an overall design developed by the System-wide Program Committee. The staff of the I & R units must be aware of this design, since they have the responsibility for making necessary changes or adaptations to provide for the needs of the students in their unit.

Objectives are set by selecting the content to be taught and by identifying the reasonable student outcomes that the IIC would expect after instruction on that content. Usually this is done by selecting a mathematics text series and referencing instructional objectives to it. Recently, many elementary texts have included lists of objectives related to their content, and some even publish tests for the objectives.

Step 2: Selecting appropriate objectives

Once the objectives in mathematics have been decided on, step 2 calls for the I & R units to identify which of the objectives are likely to be appropriate for their particular group of students. The appropriateness of the objectives cannot be precisely decided at this stage, but some rough estimate is possible. For example, children who are just beginning school would clearly not be expected to add fractions, nor (we hope) would eleven-year-olds need to learn how to count to ten. Of course, the decisions are not always so obvious, and the first "appropriate" set of objectives for a group of students may need to be revised later.

Most published mathematics materials give considerable assistance in making these first decisions. If the students were in the same program the year before, there should be achievement records for each child. Using these records of a student's past achievement, the teachers in the unit can make a better judgment about the most appropriate objectives for that student to attain in the new year. In any event, after the initial decision is made, ample opportunity to revise the first estimate will be provided when assessment begins (step 3).

Step 3: Using assessment data in planning

It is essential to assess certain information before instruction begins in order to (1) decide on the initial placement of the students, (2) check their attainment of prerequisite objectives, and (3) determine their possible mastery of the objectives of a certain topic prior to instruction in that particular topic.

Placing students in the proper instructional groups is essential. At the beginning of a new school year, the teachers may not know many of the students in their unit. To obtain much useful information about student achievement in a relatively short time, they can use placement tests. A paper-and-pencil placement test is usually designed to assess quickly a fairly large number of objectives and to give an overall picture of the student's level of knowledge and skills. From the results, the student's instructional needs can be inferred. Since children (especially younger ones) should not be given long tests and since it is likely that a large number of objectives must be covered, a placement test should not attempt to assess each objective in depth. Rather, the test should provide initial information on a student's background in mathematics. As the teacher gets to know the child, more detailed information can be added to the child's record.

In addition to assessing achievement in order to form the instructional groups, the IGE teacher must also try to determine each student's preferred learning style. There is no simple way to do this. The teacher will need to spend considerable time observing the student before drawing any conclusions. The teacher can observe whether individual students work well independently or whether they need to have contact with the teacher or other students. Some students learn very well when paired with a nonfriend of the opposite sex. Sometimes one child can learn well in small groups from the beginning, whereas another child may have to learn how to work cooperatively with others. Some students learn well from the printed page, but most find it helpful to have a chance to experiment with physical materials. One child may be satisfied in working on a simple drill exercise; another may do better when playing a game that works toward the same objective. These aspects of the ways in which students learn are best assessed over a fairly substantial length of time, probably the first month or two of the school year. And since students change, the teacher will need to be alert to these changes, noting them through the careful and systematic observation of the students on a regular basis.

The teacher must also attempt to determine each student's level of motivation. Through conferences or in the regular instructional setting, the teacher can try to determine students' attitudes toward mathematics and help them develop positive feelings toward the subject by involving them in appropriate activities.

Gathering information at the beginning of school is only one of the times when obtaining assessment data before instruction is useful. A checkup test can be used to verify students' attainment of objectives that are prerequisite for starting a new topic. These tests are especially important for those topics whose prerequisite objectives were taught some months previously; they also give information about the maintenance of these objectives. The results from this type of test will help the teacher know how to treat students as they move through the topic. If a student fails to master certain objectives, then a review of them is in order before starting work toward new objectives of the topic.

For example, suppose that students have been working on a geometry topic and the teacher decides to return to the arithmetic strand for work on the addition and subtraction algorithms for whole numbers. Before children can use the algorithm efficiently to solve problems like $346 + 279 = \square$, they first need to know the basic facts of addition. In this situation, the teacher needs to be certain that each child not only reached at some time in the past those objectives dealing with basic addition facts but also has these basic facts well in mind at the present.

In addition to using checkup tests to be sure that children have mastered the prerequisite objectives for a topic, teachers may also want to ascertain whether some students have already attained the regular objectives of a particular topic. Here a pretest of the topic's objectives may be administered before the topic is started. The results can be used to determine the extent to which individual students need exposure to the topic's activities.

Step 4: Setting objectives for individual students

Once the teacher has assessed a student's level of achievement, learning style, and level of motivation, the next step in preparing a unit of instruction is to identify a specific set of objectives for that student (step 4 in fig. 1.2).

Although at first glance it may seem most efficient to teach one objective at a time, this is usually not true. First, many objectives are so closely related that separating them would be inappropriate. For example, when students are given a set of word problems that present joining and separating situations, they are expected to be able (1) to write open addition or subtraction sentences representing the situations, (2) to solve each open sentence, and (3) to validate their solutions. Thus, for the same problem three different but related outcomes are expected.

Second, for practical reasons closely related objectives are often grouped so that students can achieve them in a reasonable amount of time and with a reasonable amount of record keeping by the teacher. For example, in *Developing Mathematical Processes* (the examples for this essay are taken primarily from *Developing Mathematical Processes,* an elementary mathematics program developed specifically for IGE [Romberg et al. 1974, 1975, 1976]), each set of the objectives that fit together in a topic usually takes about two to three weeks to teach, which seems to be a reasonable amount of time to spend before students are reassessed.

Third, since many objectives are hierarchically related, it is important to review those taught earlier and to prepare for those to be taught later as well as to instruct students to mastery (or to some other desired criterion) on those objectives composing the main emphasis of a topic or an instructional sequence. These latter can be called *regular objectives*. For example, review, regular, and preparatory objectives can be specified for many topics, as in figure 1.3. The single regular objective calls for students to write open addition and subtraction sentences with the numbers 0–99. To accomplish this, students review the objectives attained earlier involving physical and pictorial representations for such problems. But equally important, these activities are preparatory to mastering a later objective involving the addition and subtraction algorithms.

Step 5: Organizing the instructional environment

The instructional environment for teaching mathematics can be organized

Number Sentences 0–99

A variety of situations involving the numbers 0–99 are represented by sentences. The children solve open sentences about joining, separating, and difference using objects or pictures. The regrouping that occurs when the children are solving these problems is focused on as a background for algorithms (Topic 40).

Regular

1. Given an open equalizing, joining, or separating situation involving the numbers 0–99, writes a sentence that represents that situation.

Preparatory

2. Given a grouped set of objects—or its representation—and an instruction to regroup, appropriately regroups the objects.
3. Given two numbers whose sum is 0–99, computes their sum.
4. Given two numbers 0–99, computes their difference.
5. Given an open sentence involving the numbers 0–99, solves it.

Review

6. Given a number 0–99, represents it physically or pictorially.

Fig. 1.3. Regular, preparatory, and review objectives (adapted from Romberg et al. [1974, 1975, 1976, topic 38])

in a variety of ways. Using appropriate group sizes is important in IGE schools; the staff of I & R units will want to organize large groups at some times and small groups at other times, according to the activity. For some activities, students will work independently; for others, they may work best in pairs. In deciding how to organize the groups, the unit staff will want to consider how best to meet the students' needs. Sometimes the students themselves can choose whether they want to work individually. Or the teachers may decide that certain students work best independently —however, the teachers may have those same students sometimes work in a group for other reasons: for example, they may be shy and need the social contact. The unit staff should consider *all* the needs of their students in making these kinds of instructional decisions.

When students are working in small groups or independently, it is critical that the teacher be able to move about the room to respond to student needs. At this time the teacher can assess students by observation and can ask probing questions that extend understanding. If some students are having difficulty with a particular mathematical idea, this is a good time to provide the extra individual help that can get them moving again.

When students are actively engaged in learning mathematics, they should be expected to move about in a purposeful manner. Of course, this does not mean that students working in a large group under the teacher's direction should be allowed to wander off in the middle of a discussion. But students working in small groups or individually should be allowed to move about as the activity requires. They may need to ask a question of the teacher or get the advice of a friend or obtain additional supplies or materials.

Allowing verbal interaction is important, too, when students are working in small groups. They need to talk to other members of the group. This does not mean that they should talk all the time, but in many situations conversations among students are very important. They can share ideas and learn from each other in many different kinds of mathematical activities.

Arranging space, equipment, and materials for their effective use is another aspect of organization. Teachers will need to arrange the space, equipment, and materials of their unit area according to the needs of different activities. If small groups are being used, one area may be arranged with an appropriate number of tables. If tables are not available, desks can be put together to make a satisfactory working area for small groups. Another area may be arranged for a large-group activity. It may be desirable to move desks aside to use the available floor space of this area.

Another aspect of space arrangement is particularly important—one area should be designated a mathematics center and used for storing materials. Such an area is especially important for programs that use a lot of manipulative materials. Students can go there and work on puzzles or use the materials to solve problems during their free time. Space should also be provided for exhibiting graphs, projects, and other student work. Younger children in particular need a place to experiment with new manipulative materials before they are used in formal instruction in the classroom; a mathematics center thus provides an appropriate location for some preliminary investigation.

Some teachers become quite concerned at the thought of students working together in small groups, talking about the activity, and moving around the room to get materials. Many teachers have learned to provide this kind of classroom environment, but it does not happen automatically. The unit staffs need to plan activities that will help students develop their ability to work independently. Before beginning such activities, a teacher can discuss with the students what their responsibilities are, making clear what is expected of them.

With some assistance, most students will quickly adjust to their role as active learners of mathematics. As they work with others in small groups discussing problems and learning mathematics, their communication skills will improve. Often a teacher will notice that an active learner develops a more positive attitude toward mathematics. Similarly, students may notice that the teacher's view of mathematics appears to change. Perhaps everyone's view of mathematics is altered when the subject no longer is a continual succession of drill-and-practice exercises but instead provides opportunity to discover and discuss interesting ideas. These changes can result when spaces, materials, and equipment are carefully organized to provide a suitable environment for more active learning.

Learning stations are another example of how the instructional environment can be organized. After students have had experiences in graphing and in measuring distances in several standard units, the following activity using learning stations provides them with opportunities to apply what they have learned in solving real-world problems (Romberg and Montgomery 1973).

The I & R unit staff first selects a location suitable for learning stations. This could be a separate room or a section of an open area. The unit staff plans and sets up (with the help of the students) the different learning stations—perhaps as many as ten or twenty. (One way of arranging the stations in a separate room is shown in fig. 1.4.) Each station includes an

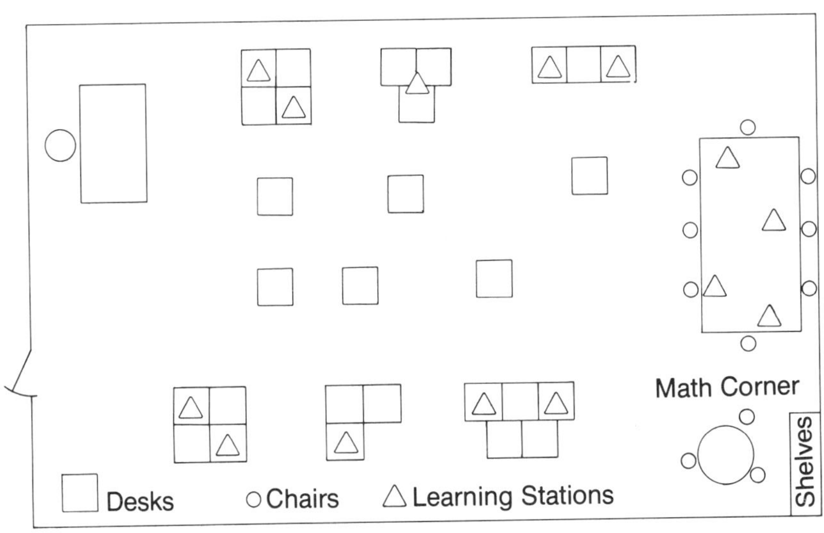

Fig. 1.4. Possible arrangement of learning stations in one room

activity card or experiment sheet and the materials needed to solve the problem on the card. For example, in order to generate and graph the measurements specified by the experiment sheet shown in figure 1.5, the learning station should be provided with a meterstick, metric ruler, wooden ball, and a piece of cardboard. Other stations would have similar sheets and manipulative materials.

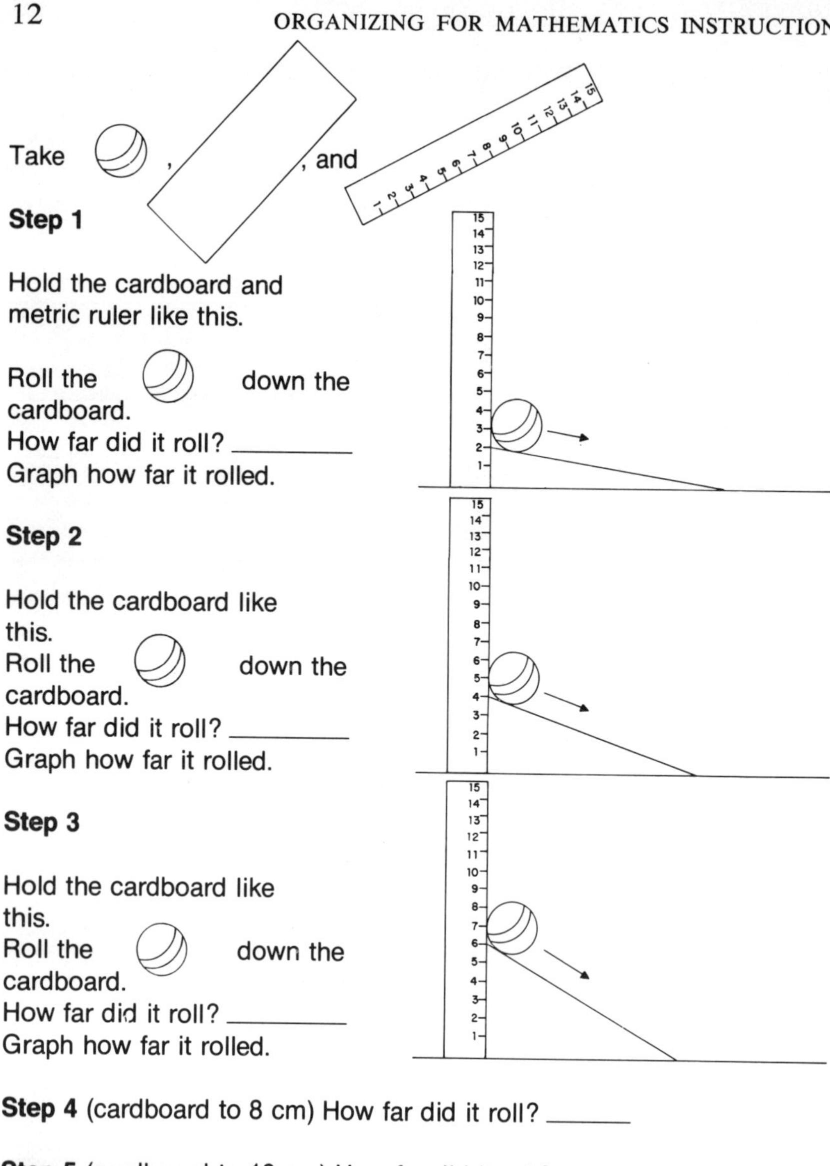

Step 1

Hold the cardboard and
metric ruler like this.

Roll the down the
cardboard.
How far did it roll? _________
Graph how far it rolled.

Step 2

Hold the cardboard like
this.
Roll the down the
cardboard.
How far did it roll? _________
Graph how far it rolled.

Step 3

Hold the cardboard like
this.
Roll the down the
cardboard.
How far did it roll? _________
Graph how far it rolled.

Step 4 (cardboard to 8 cm) How far did it roll? _______

Step 5 (cardboard to 10 cm) How far did it roll? _______

Step 6 (cardboard to 12 cm) How far did it roll? _______

Fig. 1.5. An experiment sheet

Once a learning station has been set up, the teacher discusses the
activity briefly with the children, perhaps relating it to earlier work on
graphing or measuring distances. Then individuals or groups move from
station to station, solving the problem posed at each station and recording

the results on the worksheet or graph paper provided. The teacher also moves about, helping those having difficulty and asking students to explain what their graphs represent. The teacher can thus keep track of the progress of each student or small group by observing their graphs or worksheets.

Usually, not every student will complete the work at every station. Students who finish early can help others, or the teacher may prefer to have them move on to other activities.

When the teacher feels that the students are ready to close the activity, it is sometimes helpful to bring them together to summarize what has been accomplished. This is when a teacher can ask what would happen if a longer piece of cardboard were used or if a cylinder or a marble were used instead of the ball. The teacher can also relate this activity to the objectives that the students achieved earlier. For example, their previous experience in measuring lengths can be discussed in terms of the repeated measurements necessary in the experiment in figure 1.5, and previous work in addition can be used to find the total lengths.

Activities that use learning stations are important in helping children adjust to a more active approach to mathematics. The structure of these activities is clear, and students can readily understand their responsibilities as active learners—moving to a learning station, manipulating objects, discussing problems with other students and with the teacher, and then moving on to another station.

The principles of instruction related to the organization of the classroom and the learning environment can be summarized as follows:

1. Before beginning an activity, the unit staff should organize the students in the appropriate mode—individually, in pairs, in small groups, or in large groups.

2. When children are working in small groups or individually, the teacher should move from group to group or from individual to individual, acting as a source of aid for the students.

3. The teacher should allow students to move purposefully about the room to obtain materials, to consult with others, or for other task-oriented reasons.

4. The teacher should allow students to interact verbally while working on the activity.

5. The teacher should arrange equipment and materials in a way that is suitable for the activity.

Although these principles are important, they are only a guide to what the teacher should do. Applying principles of instruction in teaching mathematics is a complex task, and teachers will need to adapt these principles to their own teaching styles. Students are complex, too; not every teaching style is appropriate for every student, and the same student can even have needs that vary from day to day. By paying attention to the special needs of each student, IGE teachers will be able to make the best decisions about how to provide instruction designed for the individual student.

Step 6: Evaluation of student progress

Once the instructional programs have been started, teachers will need to monitor student performance during the instructional sequence. Monitoring can be done by observing students and by collecting samples of their work.

Assessing students by observing them as they work on activities is an effective teaching skill. (And it does take skill to direct children in a mathematical activity while at the same time assessing them and recording their level of achievement in the objectives of the lesson.) Usually it is best to observe only about four or five students during an activity. Observations can be recorded either during the activity or, if it is more convenient, after it. By carefully observing children during instruction, the teacher can identify a student who is having difficulty and give appropriate instruction before the child goes too far astray.

Normally the teacher will want to conclude the topic by using a posttest to assess student performance on the objectives. Then, as noted in step 7 (see fig. 1.2), students who have attained the objectives are ready to go on to a new topic. Those who did not reach the objectives may need to be reassessed to determine whether they should be working on different goals (it will then be necessary to cycle back through the Instructional Programming Model) or whether they perhaps need a different type of instructional program that will take their learning style into account more fully. Decisions about these students can be complicated.

Decisions regarding the attainment of objectives are sometimes made at two or three levels for objective-referenced testing. For example, in spiral programs a second level of achievement below the mastery level is sometimes added to distinguish between those students who have nearly mastered an objective and those who still need considerable help. This is done because instructional decisions will often differ for these groups. Mastery is usually defined as performance high enough to enable the student, given a similar activity, to deal with that activity satisfactorily. This performance level is often defined in terms of a percentage of correct responses to similar test items or tasks (80 percent is a common criterion set for mastery).

When teachers observe student behavior during an activity, they must judge from their *observations* whether (1) help is needed, (2) progress is evident, or (3) the concept or skill has been mastered. However, for all the more formal assessment procedures, including tests, they should decide on *specific criteria* for judging the three levels of achievement. For example, suppose there are ten items that test a regular objective. Children who respond correctly to eight or more items receive an M (mastery); those who respond correctly to five, six, or seven items receive a P (making progress); and those who respond correctly to four or fewer receive an N (needs help). These more formal assessments are always to be used in conjunction with teacher judgment.

Assessing educational progress in a curriculum built around a specified set of measurable instructional objectives departs from the traditional ways used to determine a pupil's standing in the class. Pupil progress is measured by the number of objectives attained and the level at which they are attained. It is, in part, from a criterion level such as 80 percent correct that the phrase "criterion-referenced test" comes, as distinguished from a test that is norm referenced. One's score on a norm-referenced test indicates a level of achievement that is compared with the group at large (the "norm"). It is a measure of relative standing within the group and is communicated through grade equivalents, standard deviations above and below the mean, stanines, percentiles, and so on.

Step 7: Keeping records of assessments

After making assessments, teachers should keep records, so that the information is handy in making decisions relative to each child. Since children progress through individually guided mathematics according to their rate of attaining objectives and not by their years of school attendance, it becomes extremely important to keep track of the level of achievement for all objectives. Sometimes record-keeping devices are included in the mathematics series being used. For example, three different kinds of pupil-performance records are used in *Developing Mathematical Processes*

TOPIC ____ CHECKLIST	OBJECTIVE									
	#		#		#		#		#	
Name	Observe	T.I.	Observe	T.I.	Observe	T.I.	Observe	T.I.	Observe	T.I.
1										
2										
3										
4										
5										
6										
7										
8										

GROUP RECORD CARD
TOPICS 51-60

TOPIC		OBJECTIVE	NAME	1	2	3	4	5	6	7	8
51		(no regular objectives)									
52	1	computes sum 0-999									
	2	computes difference 0-999									
	3	solves + or − sentence 0-999									
	4	validates sentence 0-999									
53	1	describes location on grid									
	2	identifies location on grid									
54		(no regular objectives)									
55	1	states whether fraction correct									
	2	represents common fraction									

(Romberg et al. 1974): the topic checklist, the group record card, and the individual progress sheet. Although it is not necessary to use all three, each is designed to serve a special purpose.

The topic checklist is designed especially for observing students' performance during activities. The group record card is similar to the topic checklist, but it contains information on ten topics instead of one, thereby providing information about which students have completed which topics. The teacher can tell at a glance which students have mastered the prerequisites for a certain topic. Thus, the group record card is particularly useful in forming instructional groups or making other decisions about instructional arrangements. The individual progress sheet is designed for keeping substantial records on a single student. It has space for data on the achievement of objectives for ten topics; it is useful, then, for deciding when an individual has the knowledge prerequisite for a particular topic.

INDIVIDUAL PROGRESS SHEET TOPICS 51-60

Student _______________ Teacher _______________ Date _______________

TOPIC		OBJECTIVE	M	P	N
51		(no regular objectives)			
52	1	computes sum 0-999			
	2	computes difference 0-999			
	3	solves + or − sentence 0-999			
	4	validates sentence 0-999			
53	1	describes location on grid			
	2	identifies location on grid			
54		(no regular objectives)			
55	1	states whether fraction correct			
	2	represents common fraction			

M = Mastery

P = Making progress

N = Needs considerable help

Many mathematics series do not provide such a variety of record-keeping devices, but similar forms are not difficult to construct. They do require, however, a list of objectives that are properly sequenced and conveniently coded for easy reference. Also, the records should be designed to allow the information to be easily transferred from one form to another.

The precise details of the pupil-performance records are generally not crucial. What is crucial is that the unit has kept records to help plan instructional programs that are appropriate for the individual student.

Summary

At the beginning of this essay, an *effective* program was identified as one that adapts instruction for the individual student. *Efficient* instruction was defined within the IGE Instructional Programming Model and included five principles. First, an IGE mathematics program must be objective referenced. Second, planning for instruction in IGE is done by appropriate groups of professionals. Third, an extensive system of criterion-referenced assessment and record-keeping procedures is used to evaluate each student. Fourth, this information is then used to group students for differential instruction. And fifth, teachers systematically provide different activities for different groups of students.

REFERENCES

Klausmeier, Herbert. *Individually Guided Education: An Alternative System of Elementary Schooling.* New Haven, Conn.: Yale University, Center for the Study of Education, 1972.

Klausmeier, Herbert, Richard Morrow, and James Walter. *Individually Guided Education in the Multiunit School.* Madison, Wis.: Wisconsin Research and Development Center, The University of Wisconsin—Madison, 1968.

Klausmeier, Herbert, Mary Quilling, Juanita Sorenson, Russell Way, and George Glasrud. *Individually Guided Education and the Multiunit School: Guidelines for Implementation.* Madison, Wis.: Wisconsin Research and Development Center, The University of Wisconsin—Madison, 1971.

Klausmeier, Herbert, Richard Rossmiller, and Mary Saily. *Individually Guided Education: Concepts and Practices.* New York: Academic Press, 1976.

Lovell, Kenneth. *Intellectual Growth and Understanding Mathematics.* Columbus, Ohio: ERIC Information Analysis Center for Science and Mathematics Education, The University of Ohio, 1971.

Romberg, Thomas A., John G. Harvey, James M. Moser, Mary E. Montgomery, and Marcia E. Dana. *Developing Mathematical Processes.* Chicago: Rand McNally & Co., 1974, 1975, 1976.

Romberg, Thomas A., and Mary E. Montgomery. *Developing Mathematical Processes: A Different Kind of Individualized Program.* Project paper 73-1. Madison, Wis.: Wisconsin Research and Development Center, The University of Wisconsin—Madison, 1973.

2

Organizing for Individualization: A Departmental Model

Peggy A. House

The task of developing instructional alternatives can be undertaken on several levels, from large-scale, system-wide programs to the efforts of individual classroom teachers. Lying between these extremes is the option of departmental reorganization, a model for which is outlined here.

The model focuses on the secondary school (which includes junior as well as senior high school) because at that level the potential mathematics curriculum is rich and varied. Further, it is assumed that high school students can begin to accept increasing responsibility for making educational decisions in a learning environment that is supportive and nonthreatening.

The departmental model discussed here lends itself to a variety of patterns of school organization. Thus, the discussion will be developed in three parts. The first will describe the system as it applies to an open or flexibly scheduled school, since this is the setting in which it fits most naturally. The second part of the essay will attempt to present objectively the advantages as well as the limitations that can be anticipated with this system. Finally, since open schools are a minority, the third part will consider alternatives for adapting the program to those schools having traditional scheduling patterns.

Applying the Model to Flexibly Scheduled Schools

The system described in this section centralizes learning activities in a large mathematics resource area. There students and teachers of many different backgrounds work together—sometimes cooperatively, sometimes independently—on a wide variety of mathematics learning tasks. Students are actively engaged in problem solving, laboratory experiences, discussions, and research, in addition to studying printed resources and working standard exercises. This is achieved by introducing changes in a number of the aspects of curricular and instructional design, each of which must be examined further.

Changing teacher and student roles

The first significant change necessitated by this model is in the roles of both teachers and students. Since the emphasis is on active learning by the student, there is a concomitant emphasis on one-to-one and small-group interaction. Several teachers are available for individual consultation, peer teaching is actively promoted, and the students are encouraged to learn and to study with others. Much of the teacher's activity is shifted from that of an information-giver to that of a resource person. Yet teachers can easily arrange for more structured presentations to groups of varied sizes whenever the needs of the students or the nature of the subject matter make that desirable. These elements of role change become clearer in the following paragraphs, which outline other components of the model.

Reorganizing places and spaces

The learning environment contributes significantly to the success of this system. Ideally there should be a large, centralized learning place or a cluster of adjoining rooms (for example, an unused cafeteria, study hall, or all-purpose room; a pod; or a set of adjacent rooms with the connecting corridor blocked off and converted to a study area). Each teacher in the department has an identifiable station, but all teachers and students function in this common space. As a result, students can interact with other learners of diverse backgrounds, and more advanced students can easily provide assistance to their peers. Also, students have access to a wide variety of teachers in addition to the one with whom they are currently studying. The atmosphere in the central space accommodates independent work, group study, and teacher-student or student-student consultation. A conversational level of noise is common.

Adjacent to the center should be smaller spaces for more specialized use. These should include a small classroom for seminars, lectures, films, or other group presentations; a computing center; a quiet study area; game and activity rooms; an area for collections of materials; a resource center; faculty office space; and a testing center (unless this is accommodated elsewhere in the school).

Organizing time

Time is one of the most variable elements in the system. Within the flexibly scheduled or open school, students select both the amount of time spent in the mathematics center and its distribution in the school day or week. To accommodate those students who may need more assistance in managing their time, provisions can be made to require these individuals to sign in for an appropriate minimum attendance each day or week. It is also advantageous to have a defined period of time—for example, Mondays between 10:00 and 11:00—designated the "math hour." During this hour, mathematics-department activities take precedence over all other school functions so that it is possible to convene various groups of students (for example, all algebra students or all sophomores) for purposes such as large-group presentations or group testing whenever these activities are appropriate. Of course, other departments will also probably request a similar type of hour—such as an English hour, a social studies hour, and so on.

Several examples of possible schedules are shown in figures 2.1 through 2.3. Anne attends a modularly scheduled school, and her schedule is shown in figure 2.1. In this school, mods 5, 6, and 7 on Mondays are reserved for the math hour. The English, science, French, and social studies classes hold regularly scheduled large- or small-group meetings as shown. By her own choice, Anne has decided to schedule herself into the mathematics center on Wednesday and Friday mornings (Monday, Wednesday, and Friday on those weeks when the math-hour option is not exercised) and on Tuesday and Thursday afternoons. For her, attendance in the center is optional; hence, she sometimes skips a session when work in other subjects is pressing (such as on the day of a major social studies test). At other times she spends more time in the center in order to complete a unit, do a laboratory exercise, work on problems with a fellow student, or discuss difficulties with a teacher. Anne is able to do this, since her unscheduled mods can be spent in the library, study hall, mathematics center, or other resource centers.

Sara, who attends the same school, had a schedule similar to Anne's. However, Sara is less motivated than Anne and finds it difficult to complete the required credits in mathematics. At the beginning of the year she, too, planned her own schedule of attendance in the mathematics center, but when she did not follow it, her mathematics achievement declined sharply.

Student's Schedule

Anne

Mod	Monday	Tuesday	Wednesday	Thursday	Friday
1	English	Science	English	Science	English
2	”	”	”	”	”
3		”		”	
4		”		”	
5	Math Hour		✕	”	✕
6	” ”			”	
7	” ”				
8	Lunch	Lunch	Lunch	Lunch	Lunch
9	Soc. Stud.		Soc. Stud.		Soc. Stud.
10	” ”	✕	” ”	✕	” ”
11					
12					
13	French		French		French
14	”	Phys. Ed.	”	Phys. Ed.	”
15		” ”		” ”	
16		” ”		” ”	

✕ = self-scheduled time spent in mathematics center

Fig. 2.1

After a conference with her teacher, both agreed to a required attendance schedule under which Sara must sign in at the center during mods 5 through 7 every day. She and the teacher reevaluate her need for required attendance at the end of each month.

Tom takes the same subjects Anne does, but Tom attends an open school in which students must schedule their own time in all subjects, not just in mathematics. The only fixed points in Tom's schedule (fig. 2.2) are the hours reserved for the various departments. Tom, like Anne, can schedule his own pattern of attendance in the mathematics center. Although he has more flexibility than Anne in the time available for mathematics, he also has the greater responsibility of balancing his time among the five subjects.

Paul Cooper is a mathematics teacher who might be working in either Anne's or Tom's school. His schedule is shown in figure 2.3. Because he has a large number of students enrolled in the basic algebra and geometry courses, he has reserved the early morning mods as indicated, and the students in those courses are free to attend on an optional basis. In this way he is able to work with groups of students who are encountering common difficulties. He also uses these mods on occasion to present related topics for background or enrichment. Like the students, Mr. Cooper must reserve the Monday math hour; in addition, he has a weekly staff meeting

| | | | Student's Schedule | | |
| | | | Tom | | |
Mod	Monday	Tuesday	Wednesday	Thursday	Friday
1			Science Hour		
2			" "		
3			" "		
4					
5	Math Hour				
6	" "				
7	" "				
8					French Hour
9					" "
10					" "
11					
12		Soc. Stud.			
13		Hour			
14		"		English Hour	
15				" "	
16				" "	

Fig. 2.2

with other members of the mathematics department and a Friday afternoon seminar in which advanced mathematics students pursue topics of interest with faculty members. Finally, he has set aside the last three mods of the

| | | | Teacher's Schedule | | |
| | | | Mr. Cooper | | |
Mod	Monday	Tuesday	Wednesday	Thursday	Friday
1	Algebra	Geometry		Algebra	Geometry
2	"	"		"	"
3					
4					
5	Math Hour				
6	" "				
7	" "				
8					
9					Seminar
10			Department		"
11			Meeting		"
12					
13					
14	Appointments	Appointments	Appointments	Appointments	Appointments
15	" "	" "	" "	" "	" "
16	" "	" "	" "	" "	" "

Fig. 2.3

day for work with individuals or small groups who have made previous appointments. The rest of the time he is in the center to assist students on a walk-in basis.

Another aspect of time flexibility stems from the adoption of a mastery learning model in which credit is granted when the student demonstrates competence on certain prespecified objectives. (For a more complete discussion of mastery learning, see *Mastery Learning, Theory and Practice,* edited by James H. Block [New York: Holt, Rinehart & Winston, 1971].) Because of the greater accommodation of individual needs under this arrangement, students are able to take whatever time is needed to attain mastery either by extending the number of days they study on a unit or by increasing the amount of time they spend each day on mathematics. Some aspects of the record keeping involved under this system will be discussed later.

Restructuring the curriculum

The flexibility and increased options provided by the model discussed in this essay are derived principally from breaking down the standard curriculum into many short-term units, which eliminate nonessential sequencing and greatly increase learning alternatives. This reorganization of the curriculum can readily be achieved without making major changes in content and without requiring curricular decisions beyond the usual decision-making powers of the classroom teacher, although the curriculum is enhanced when teachers are able to exercise significant choices.

Take geometry, for example. Instead of the usual year-long sequence, a short basic-concepts unit would form the core course. An example of the content this basic unit might cover is given in figure 2.4. The content is further extended by a set of units that might include such topics as similarity, congruences, properties of polygons, right triangles, quadrilaterals, special polygons, circles, conic sections, polyhedra, geometric proofs, constructions, loci, coordinate geometry, solid geometry, and so forth—topics already considered part of the standard curriculum. In addition, other units might focus on transformational geometry, symmetry, paper folding and curve stitching, geometry in art and architecture, scale drawings, vector geometry, polar coordinates, projections, affine geometry, advanced concepts of Euclidean geometry, non-Euclidean geometries, geometry in N-space, the golden section, tessellations, history of geometry, applications of geometry, and more.

Outlines of each unit are prepared and distributed to the student as a means of specifying mastery criteria (objectives), preassessments, suggested learning activities (readings, films, problems, worksheets, laboratory experiments, etc.), assignments, and self-evaluations. These units can take

---

Introduction to Geometry

Core Unit—Fundamental Concepts

Properties of points, lines, planes, and their subsets
Relationships of lines: intersections, angles, parallels, perpendiculars
Geometric equalities and inequalities
Measures of angles, arcs, distances, area, perimeter
Congruent triangles
Similar triangles
Properties of special triangles (right, equilateral, isosceles)
Quadrilaterals
Circles
Regular polygons
Solid figures; volume

Fig. 2.4

many different forms. Some may guide the student through a single standard text. Others may direct the student to use a variety of texts, readings, programmed materials, or other resources. Some may be coordinated with audiotapes or videotapes, films, film loops, and so on. Still others may include self-contained instructional materials probably written by the teachers. A few will be exploratory and very open-ended. Often students will be able to work through units either independently or with others. At other times, the nature of the unit may suggest that students be required to enroll in groups of, say, six to eight. For example, the unit on the history

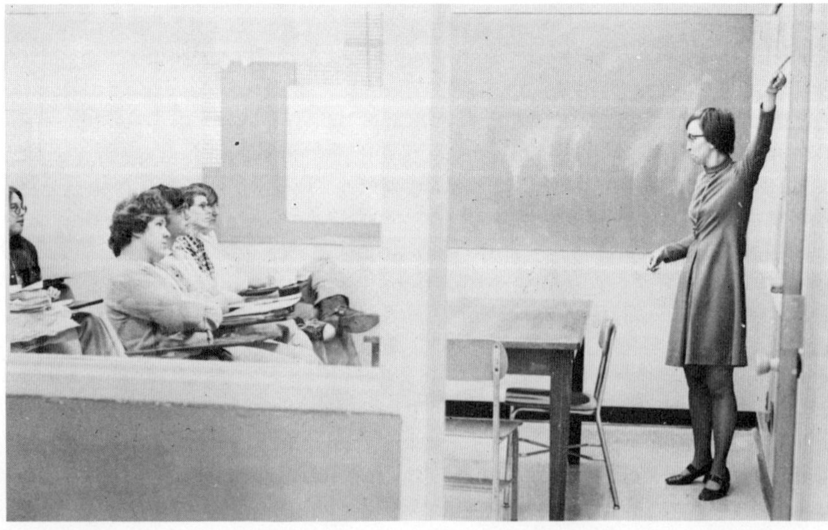

of geometry might be designed as a seminar, which would presume such a group, whereas the unit on geometric constructions could easily be done independently.

A helpful technique for increasing student interaction is to post in the mathematics center the names and homerooms of students currently enrolled in each course. This helps students identify possible learning partners and enables them to plan and work cooperatively. For required introductory courses or courses in which a large number of students enroll, it is often advantageous to engage the more advanced students as volunteer tutors and to set up and publicize a regular schedule of the times when these tutors will be available in the center.

For allocating credit, the standard can be taken as ten semester hours, or ten credits, for the typical year course (5 hours a week each semester $\times$ 2 semesters). Since the usual school term is approximately thirty-six weeks, this results in close to 180 hours of classroom instruction, or approximately 18 hours for each credit (not counting time spent on homework or out-of-class study). This formula can serve as a rough guide to allotting credit for units. Most units will be expected to offer from one to three credits. In the example from geometry, the core unit (fig. 2.4) would warrant at least three credits, perhaps more; the majority of the other units would have one credit each. Note that the suggested list alone would produce more than triple the typical ten-credit geometry offering. Students might be required to take, say, eight credits of geometry: a three-credit core unit plus five other credits (units) of their choice, selected with a teacher's guidance. Another possibility is illustrated in figure 2.5. Here a student might be required to elect at least seven units beyond the core: four from group A, two from group B, and one from group C. Many other variations are possible, and it is important to keep in mind that students always have the option of taking more than the required number of credits.

Geometry Electives

Group A	Group B	Group C
Similarity	Vector geometry	History of geometry
Congruence	Transformational	Tessellations
Properties of polygons	geometry	The golden section
Right triangles	Loci	Affine geometry
Quadrilaterals	Polyhedra	Scale drawings
Circles	Solid geometry	Geometry in art and
Special polygons	Applications of	architecture
Coordinate geometry	geometry	Paper folding and
Geometric proofs	Symmetry	curve stitching
Conic sections	Geometric constructions	Geometry of N-dimensions
	Polar coordinates	Projections
	Advanced Euclidean	Non-Euclidean geometry
	geometry	

Fig. 2.5

It is important to avoid all nonessential sequencing when designing such courses. Unless this is done, the flexibility that the curriculum design is intended to increase will be short-circuited. However, when the task of modularizing the curriculum is undertaken as suggested above, it is often surprising how much presumably sequential material is found to be, in fact, quite flexible. Where sequencing cannot be avoided, those units will specify prerequisites.

Ideally, students should be free to select the teachers with whom they will enroll. It would not be necessary to take all the geometry units from the same instructor, although it would not be uncommon to do so. To make the total curriculum more manageable, the mathematics faculty may choose to designate some subset of itself as instructors for certain courses, particularly in more specialized courses or in units for which some teachers have a background of interest and experience. Teachers must also have the opportunity to turn down a student if their student load is currently excessive. The student can then either seek another instructor or wait until the chosen teacher is available.

Management

Classroom management, record keeping and reporting, student accountability, effective use of time, and the availability of materials and resources are concerns of every teacher and are crucial to the success of the system.

Classroom management must be interpreted in accordance with the goals of the program. Since these goals stress the active involvement of the students, the classroom atmosphere should encourage activity, movement, and personal interaction. If the learning center has adequate acoustical qualities, normal conversation should not disturb anyone. Purposeful movement about the room should be permitted, but it may be advantageous to restrict the time for entering and leaving the center to those times when students change classes elsewhere in the school (usually a matter of administrative convenience). A recommended guideline is that reasonable noise and movement are in order when they contribute to a student's learning and when they do not infringe on the learning activities of others.

Record keeping, too, must be adapted to fit the system. Essentially this amounts to keeping track of each student's progress and monitoring the number and distribution of credits earned. Specifics for record keeping are best determined by the staff who must use them, but there should be reasonable standardization within the school mathematics department, since students are likely to enroll for courses with several different teachers.

It is hoped that the administration will support a system of credit reporting whereby the instructor can forward to the office a credit report (including the student's name, course name or number, credits earned, date completed) whenever a student has earned credit in a given unit. One copy of this report is given to the student, and one is filed in the department office. However, if the administration does not accommodate such a plan, the reporting of credits earned is merely delayed, so that such reports correspond to specified reporting periods: monthly, quarterly, semiannually, or the like.

The students' use of time is also an important concern of management, especially at the time of the departmental transition to this type of system or when a student is new to the alternative of open education. Several strategies can help students develop responsibility for using their time wisely, and instructors should be open to exploring these strategies as needed. For instance, it might be wise at first to schedule students for a minimal amount of required time in the mathematics center, perhaps three periods a week. The requirement can be lifted when the student completes the first short course. Or, as with Sara earlier, an appropriate attendance requirement can be specified if a student does not demonstrate satisfactory progress in mathematics. Some individuals, who prefer a more structured style of learning, may need a schedule of periodic meetings with the teacher. These meetings might also include some teacher-centered lessons with small groups of students. Mr. Cooper, for instance, might require such students to attend his Monday and Thursday morning algebra sessions where he will make presentations, lead discussions or problem sessions, or work individually with them. All such alternatives are legitimate and frequently desirable, but teachers are cautioned to remember that the goal is to develop increasingly more independence and responsibility in the learner. When a student no longer needs the external control of required attendance, the requirement should be eased.

To achieve learner independence, it is essential that adequate resources and materials be readily accessible to the student. Students are not likely to learn mathematics by reading it on their own any more than they do by listening to a teacher talk about mathematics. They need both reading and listening, but they also need more: they need *active involvement* with mathematics. Individually or in groups, with or without the teacher, students should be encouraged to engage in laboratory activities that help them develop mathematical concepts and understandings. But the only way such activities can be useful is to have them available when needed.

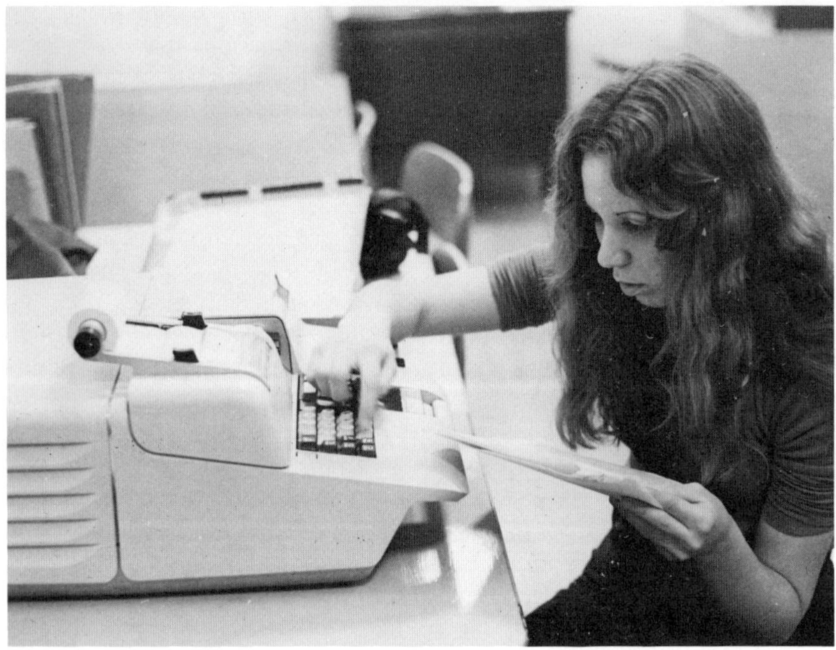

For example, an algebra student studying the concept of functions might experiment with pendulums, inclined planes, springs, shadows, bouncing balls, or any of dozens of other embodiments of functional relationships. A geometry student might investigate concepts of area through open exploration or guided discovery using a geoboard. Some might construct a three-dimensional model to demonstrate their understanding of a principle. Other students might seek to develop various problem-solving skills through certain games of strategy, or they might design a new game to illustrate a result from their study of game theory.

As the foregoing activities suggest, materials to be kept on hand include paper, tape, glue, scissors, wood, string, manipulative devices, games, graph paper, compasses, measuring devices, and so forth. The students must have ready access to the materials needed. If materials are stored in close proximity to the center and if they are adequately displayed, cataloged, and labeled, students can help themselves with a minimum of dependence on the teacher. Materials that are locked in cupboards or take two days to find are of no use to anyone.

Teachers must also keep in mind that students will need adequate direction in the use of such materials. Their presence alone does not guarantee their effective use. If concepts of area are to be developed using a geoboard, the student will need both the geoboard and the guidance for using it appropriately. Such guidance can come through verbal interaction with the teacher ("Why don't you try—?" "What would happen if—?" "Suppose you increased—") or through worksheets or printed guides ("Use your geoboard to find answers to the following questions"). If the student is to view a certain film at a given point in the development of a concept, it is important that the unit guide specify, "Before proceeding, view the film *Surveying Techniques*. The film can be checked out any time in the audiovisual room."

Evaluation

Evaluation, which is crucial, is not to be confused with grading, which may or may not be employed under this model. Evaluation involves, first, a diagnosis of the student's readiness for a unit and of his or her mastery of prerequisite skills or concepts. This is usually done by administering a pretest, which may be checked either by the teacher or by the student. If the student does not demonstrate entry-level mastery, the teacher prescribes appropriate learning activities until the student demonstrates the expected readiness. Pretests or other diagnostic instruments do not weigh in the final evaluation of achievement.

Preassessments should also give a preliminary measure of the objectives of the unit, for it is possible that a student may have already achieved the desired mastery of all or some of those objectives. If a student's performance on the pretest suggests that this may be so, he or she should be given the opportunity to demonstrate that mastery immediately. If the subsequent performance meets the mastery criteria, credit should be given, and the student can move on to another unit. No one should be expected to "learn again" what one already knows.

Consider Virginia, a student who elects to enroll in an introductory statistics course. The pretest for such a course would measure her ability to perform prerequisite arithmetic operations, including squaring a number, computing a square root or locating it in a table, substituting a value for the variable in an open sentence, and performing calculations in the correct order of operations. All these might be considered entry-level requirements, and if Virginia has not mastered some or all of them, she would be expected to do so either before beginning the unit or at the appropriate point within the unit where such mastery becomes necessary. But the pretest would also include some items corresponding to the objectives of the unit. Suppose, then, that the results of the pretest further suggest that Virginia has already learned the meaning and methods of computing measures of central tendency (mean, median) but that she is not familiar with measures of dispersion (range, standard deviation). A further check of the objective that the student be able to compute the mean and median for a set of data might then be in order. If that check confirms Virginia's mastery of the objective, she would skip those learning activities designed to achieve the central tendency objective and move directly into those activities that develop the concept of measures of dispersion. If it should happen that Virginia has, in fact, already mastered all the objectives for the introductory statistics unit, she could move immediately to the intermediate statistics course. Or, of course, she could decide that her mastery of statistics was adequate for her present need and so elect another area of mathematics entirely.

Since students work much more independently in this system, they have a much greater reliance on self-evaluation, and frequent and immediate feedback is important. Solution keys to exercises can be distributed with the unit guides or filed for student use in the resource center. Self-tests can be handled in a similar fashion. Quizzes that are required at periodic intervals in each unit are filed in the testing center and available when the student is ready. These quizzes provide two important dimensions of formative evaluation: they indicate to both student and teacher the individual student's progress toward the mastery of desired outcomes, and they give the teacher information on the effectiveness of the program, the types of difficulties experienced by students, needed revision or expansion, and so on.

Evaluation should occur when the student is ready for it. One approach is for individual students to report to the teacher when they have completed required assignments, passed required quizzes, or otherwise achieved prescribed en route objectives. The teacher then issues a test authorization slip, which the student presents at the testing center for the final, summative evaluation. Tests and quizzes are criterion referenced and specify in advance the acceptable level of mastery. Students who do not demonstrate mastery are guided to appropriate learning experiences that help them recycle through those parts of the unit not yet mastered until they can perform on the criterion evaluation at the prescribed level. Since tests taken in the testing center are returned to the teacher for correction, prescribed remediation can be indicated at the same time that the teacher communicates the test results to the student. If mastery has been indicated, the teacher issues the credit report.

As implied above, several alternative forms of quizzes and criterion tests are needed. Often these can easily be generated, at least in part, by a short computer program. Suppose, for example, that the unit in question concerns quadratic equations and that a quiz on finding the roots of quadratics is desired. By generating random triples (a,b,c), the computer can easily print out any number of quadratic equations for such quizzes. Indeed, each student can have a unique quiz if desired, although four or five forms would probably be ample to assure a different version of the quiz on second or subsequent testings. If the quiz is concerned with higher-level objectives involving applications, derivations, proofs, heuristic problem solving, or the like, teachers would construct parallel test forms in the same way they construct different tests for different sections of the same course: by locating examples in which the same objective is embodied in different, but mathematically equivalent, specific situations.

In addition, teachers soon find that informal evaluation, which occurs spontaneously in an atmosphere of heightened personal interacting, plays an important role in the mathematics program. Because teachers have much more individual contact with students, they frequently employ oral

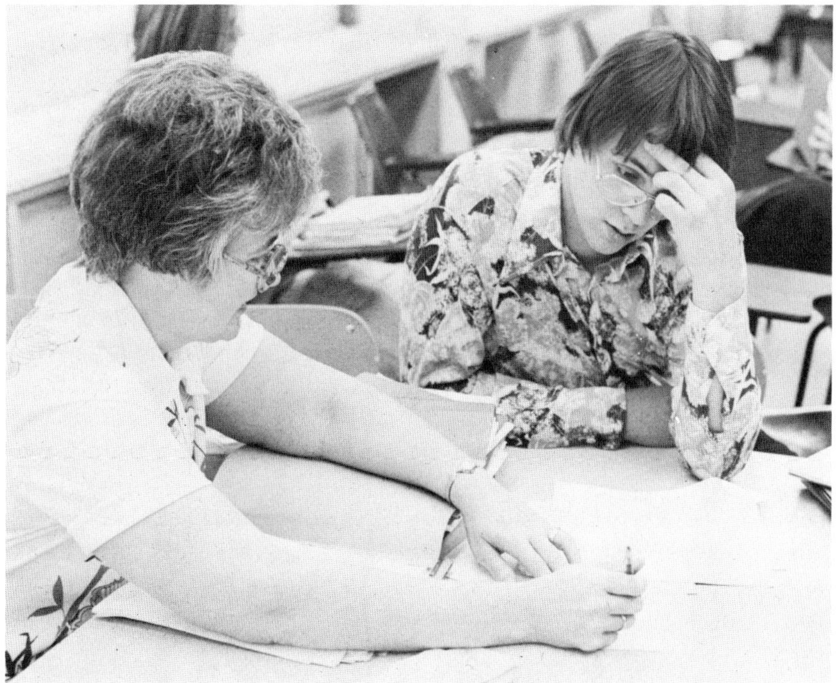

quizzes, interviews, observations of students explaining concepts to other students, or other on-the-spot measures to assess student performance.

Grades are unnecessary in a mastery environment where acceptable performance criteria are specified in advance and what is assessed is the achievement or nonachievement of mastery. What is recorded is achievement when it occurs. If, however, teachers, administrators, students, and parents are unwilling to dispense with grades, a grading system can be worked out, provided it is criterion referenced.

There is no place for norm-referenced grading in a mastery learning model. Grading is norm referenced whenever students are compared to each other, as in the traditional grading system in which, for example, an A indicates superior performance relative to others in the norm group or class, even though an A in one algebra class may correspond to the same score as a B did in another class. In a criterion-referenced system, an A represents the mastery of a prespecified set of objectives differing in quantity or level of cognitive performance from the objectives required for a B. There is no concern for normal distributions, and if all the students in a group earn an A, all will receive an A. Criterion-referenced grading is the only grading compatible with this system.

Advantages and Disadvantages

The advantages that can be anticipated from this system are those related to increasing the opportunities for student success through mastery learning and through a variety of learning alternatives.

Advantages of mastery learning

The theoretical underpinnings of mastery learning cannot be discussed here, but certain advantages should be recognized. One advantage of mastery learning is that achievement becomes a constant, whereas time becomes a variable. Thus students can experience more success in mathematics because they are not faced with new learning tasks before they are ready. Another advantage is that because of the greatly increased individualization, levels of mastery can more easily be adjusted, thus making them appropriate for learners with varied characteristics or goals.

Advantages of increased access to learning opportunities

The approach described in this essay extends to the learner an invitation to select from among many options; in addition, it enables the learner to pursue those options by increasing his or her access to a number of essential components of learning. (For a discussion of the concept of open access, see *The Open Access Curriculum,* by L. Craig Wilson [Boston: Allyn & Bacon, 1971].)

Access to the curriculum is greatly enhanced by the short-course format described earlier. This arrangement permits vast new areas of mathematics to be explored through short courses designed to introduce topics of growing contemporary significance, such as statistics or computer programming, or to make available mathematics not traditionally included in the secondary school curriculum. Because nonessential sequencing is removed, the configuration of courses can vary greatly, thus enabling the student to gain access to the curriculum through a large number of entry points at nearly any time during the high school career. Further, the short courses permit the student to elect additional study in an area of interest without making it necessary to enroll for an entire year or semester. Finally, since curricular choices need not be group decisions, the individual has greater opportunities to access the curriculum in ways compatible with his or her personal goals.

A second important increase is in the access to time. Students are no longer restricted to fixed periods of mathematics learning within the weekly school cycle with an increase in time coming only through the addition of more weeks of study; rather, they are free to spend as much or as little time as necessary in the center. This is qualitatively different from the idea of simply using more study time for mathematics, since here the student has available resources and teachers not present in the typical study hall.

A third dimension is greater access to helpers and learning partners. It is a popular misconception to assume that indvidualized instructional programs must lead inevitably to lonely learning wherein the student is isolated by a carrel, a tape recorder, or a learning packet. Rather, learning is enhanced through human interaction. In the learning center, learners have continuous access to a variety of helpers—other learners as well as teachers—since peer-group learning and student tutoring are regular and important activities that benefit not only the learner but also the helper.

Access to materials and resources is important in any educational design, and this need increases inversely to the decreasing role of the teacher as the primary source of information. The present design assumes that printed resources (texts, worksheets, programmed materials, reference books, tables, supplementary resources, etc.), audiovisual materials (films, film strips, transparencies, audiotapes or videotapes, film loops, and the respective projectors), laboratory equipment, games, models, manipulative devices, computing facilities, calculators, and any other relevant hardware or software are readily available to the students and in the closest possible proximity to the learning center. As a result, students have access to a wide variety of learning options, including the very important dimensions of "enactive" (i.e., manipulative) and "iconic" (i.e., pictorial) as well as symbolic (abstract) learning experiences. Furthermore, the individualized nature of the program and the variability of both time and learner goals frequently obviate the need for classroom quantities of some materials, and as a result, funds can instead be used to provide greater variety in resources.

In terms of student progress, the evaluation system increases the learner's access to quick feedback on his or her achievement in relation to stated goals and thereby further increases the learner's access to appropriate goals and the means of achieving them. At the same time, students who have been freed from the pressures of competition and who have learned how to study in a criterion-referenced environment usually improve notably in their ability for self-evaluation.

Teachers who implement this model of organizing for instruction will soon find other ways to increase the student's access to learning opportunities according to the unique characteristics of their particular situations. For instance, outside resource persons, out-of-school learning experiences, or courses offered at local colleges are a few of the possibilities that can easily be accommodated under this design.

Advantages of increased options for the learner

The purpose of increasing the student's access to all the preceding opportunities is to enable a concomitant increase in the learner's options, for such an increase in options is both a major reason for, and a necessary condition of, individualization.

In particular, this model allows more options for desired learning outcomes both in terms of the content of those outcomes and in terms of the level of mastery desired. The student planning further study in mathematics or a related field at the college level can project as a goal the mastery of high school trigonometry and analytic geometry or even of first-year calculus. Another student, also college bound, might elect only introductory units in trigonometry and analytic geometry in preparation for later enrollment in the college's precalculus course. A third student, not interested in higher education, might select objectives dealing with topics of consumer mathematics, measurement, or career-related applications of mathematics. All three would be required to demonstrate mastery, although the criteria for mastery would be different for each individual.

Students also have the opportunity to assume teaching roles, to learn with others or alone, to change learning partners, to vary their learning style, and to distribute their time for learning in a wide variety of ways. In short, they accept responsibility for a multitude of learning decisions according to their personal maturity and their readiness to accept the implied accountability.

Advantages of increased interaction in the classroom

Learning is an interactive process, and the model considered here is characterized by a variety of significant interactions. The interaction between teachers and students, between students and students, and between students and the curriculum have already been discussed. But an additional feature should also be emphasized: these interaction patterns are built on cooperation rather than on competition. Education is not a zero-sum game in which one party wins only if the other loses. Students should be free to share their knowledge, not forced to guard it jealously lest a classmate achieve a higher test score, which might alter the grading curve. In the criterion-referenced environment this is not possible because grades are not relative. Students are free to help each other without the fear of personal loss; indeed, they usually discover that helping another is of benefit to both.

No system is without limitations, and it would be unrealistic to pretend that this one is an exception. However, limitations usually can be managed, provided they are recognized and honestly assessed. It is appropriate, then, to examine some of the present constraints, which can be grouped into three categories: limitations related to personnel, to curriculum, and to management.

Limitations with personnel

A precondition for success is the belief by the departmental staff that, at the very least, the system has potential and represents an alternative worth trying. Users who convince themselves in advance that the system cannot work will inevitably produce a self-fulfilling prophecy. Further, it must be recognized that the system calls for some radical changes in teacher roles that will demand more, not less, of the teacher.

Students, too, must assume new roles and must learn how to learn within such a system. The biggest danger lies in the tendency for those students who function well in this environment to consume a disproportionate share of the teacher's time while the less eager fall deeper and deeper into frustration. In fairness, it must be recognized that these same students would doubtless be frustrated in a traditional classroom as well, and they are not necessarily more numerous in this system—only more visible. Some, particularly in a setting where attendance is left to the discretion of the student, will likely become missing in action. A periodic scan of teacher's records should alert the department to the identity of these MIAs who, no doubt, are in need of special help or attendance requirements.

Parents and administrators, like students and teachers, must be well informed about the goals and operational procedures of the system. In particular, parents must understand the evaluation system, the meaning of credits, and ways in which they can support, encourage, and help their children. They also must be assured that the program can benefit the students and that it will not jeopardize their later success in college or career.

Finally, the system will benefit significantly from the services of competent aides and paraprofessionals. Among the services they can render are administering the testing center, supervising the resource center and other special areas, tutoring, maintaining records, and performing clerical tasks. Often these positions can be filled by students, interns, or preservice teachers.

Limitations with curriculum

Getting started can be the biggest obstacle in planning a curriculum, and it is wise to keep the undertaking within manageable proportions. The most likely approach is to proceed in an incremental fashion by concentrating first on reorganizing the courses already in the mathematics curriculum and later adding new components.

One might decide to begin the revision with geometry. The first step would be to design the core unit, such as that suggested earlier (see fig. 2.4). Next, one would concentrate on the units in group A (fig. 2.5). Only when these are satisfactorily developed and operational would one approach the construction of topics in groups B and C. A parallel evolution would take place in the algebra and general mathematics curricula. Such developments could begin concurrently in two or three subject areas (geometry, algebra, general mathematics, for example), probably with different members of the mathematics faculty concentrating on each, or the entire staff could focus its efforts on one area the first year and introduce a second area the next year.

It is important that the curriculum be regularly updated, and revising a course is almost always easier than its initial development. One recommendation for facilitating revision is to produce only as many unit guides as the department expects to use in one year. Revisions based on evaluations of the unit's effectiveness are then made before the next printing occurs.

As the curriculum is developed, the proliferation of available courses may persent students with problems of having too many choices. It is a good idea to prepare a curriculum guide that informs students of the nature and content of courses; their prerequisites, if any; and suggestions about who might likely consider each specific course.

Teachers, too, must be aware of the dangers of overchoice. For example, if a student is learning to calculate very large and very small distances by using techniques of indirect measurement and if ten possible activities are available to the students, teachers may be tempted to assume that all ten must be completed. If a book and a film loop each develop the law of cosines equally well, teachers may still insist that the student read the book *and* view the film, even when one is sufficient. When thirty geometry units are available in the curriculum, there is the danger of distorting one's perspective by concentrating, not on the ten units a student completed well, but on the twenty units not selected.

Teachers are further cautioned to avoid the danger of producing courses that depend too much on a kind of programmed, isolated learning. Although the independence and individualization of the learner are prime goals, these must not be mistaken for complete self-instruction. Courses should include assignments and learning tasks that encourage the student to engage in discussions with the teacher or with other students. Such tasks might include laboratory activities and group as well as individual problem solving, or they might suggest that the student explain or demonstrate certain concepts to another student.

In a related consideration, teachers are also cautioned not to shortchange problem solving. It is easy to assign exercises in which the student must employ known algorithms in a reproductive, convergent style of thinking. It is much more difficult to challenge students with nonstandard problems that encourage divergent, creative thinking and that require heuristic behavior for their solution. Because of the importance of problem solving in mathematics education, teachers are encouraged to consider requiring problem-solving seminars for all mathematics students.

Limitations in management

Some of the major problems of management have already been discussed in connection with the student's use of time and the development of curriculum. Another concerns the teacher's time.

There is no ready answer to the question of what constitutes a teacher's load. An initial definition might equate it with the number of students who would be assigned to the teacher under the traditional system. Later the department could decide to assign weighted values to different courses according to the amount of teacher input required for each student. Since students determine their own pace for beginning and completing courses, a teacher's load will fluctuate throughout the year. As noted earlier, there may be occasions when a teacher is unable to accept new students until some later date.

Another constraint on management may result from a lack of available aides. In this event, departments might consider adopting a restricted schedule for the testing center, resource center, and so on. (As an example, testing might be available during the first and last hours of each day, or perhaps any time on Tuesdays and Fridays; the resource center might be open mornings only.) In the case of materials, teachers almost never consider available resources to be adequate. However, as new acquisitions are made, quantity should give way to variety, for classroom sets are seldom, if ever, needed.

The improvement of any program depends on evaluation. Adequate plans for the ongoing evaluation of students, teachers, curriculum, and instructional design are important from the outset, but care should be taken that premature decisions do not lead to the unwarranted abortion of the program. Many aspects of the system will be new to both teachers and students, and all involved will need sufficient time to adjust to changes and the freedom to explore alternatives without undue pressure or the fear of consequences.

Adapting the Model to Traditionally Scheduled Schools

So far the discussion of the proposed model has assumed an open or flexibly scheduled pattern of school organization. Although this is the setting in which the model is most easily implanted, it is not the only possibility. Many elements of the system can be assimilated into traditional patterns of school scheduling. It is always assumed that the curricular organization (i.e., short-course format) and criterion-referenced evaluation system (mastery learning) are essentially as described in the preceding sections. The changes occur only in the scheduling patterns and in some aspects of the learning environment.

Perhaps the modular schedule presents the least difficulty because it usually provides students with relatively large amounts of unscheduled time. Thus, if the physical environment and instructional design of the learning center can be arranged as previously described, the only essential change is that students must be present in the center during scheduled mathematics modules and may, in addition, spend as much unscheduled time as desired in the mathematics area. The accessibility of different teachers and other learners would only be somewhat restricted according to the schedules of those others.

Even with the less flexible six- or seven-period schedule, much of the model can be retained if the school adopts the mathematics learning center arrangement. For example, all students and teachers who are scheduled for mathematics during fourth period of the day would be in the center at that time. Although the schedule would require that they remain in the mathematics area for the entire fifty- or sixty-minute period, while there they could function as described earlier. However, students probably would not have the option of spending additional time in the center unless they have a study period, during which they might elect to return to the mathematics area instead of reporting to the study hall.

There are advantages and disadvantages to both of those situations. For those students who go to mathematics class with something less than wild enthusiasm, having a regular schedule can be a positive factor. Also, since one can predict which students will be present during a given period, it is sometimes easier to develop working groups among students currently studying similar topics of mathematics.

Because students would be scheduled into classes, one might expect, for example, that a first-period group could be composed of sophomores scheduled for geometry. (Also present in the center during first period might be another geometry group, an algebra group, and a trigonometry group.) Those sophomores, then, would all begin the year with the geometry core unit, but they would later proceed into individual patterns in terms of the number and sequence of other geometry units. Experience shows, however, that thirty students are not likely to be doing thirty different things at any given time; rather, they are more likely to be doing five or six different things. Thus, when a number of geometry students are regularly in the center during the same period, it is considerably easier for the teacher to work with small groups and thereby to avoid some of the tedium of repeating the same explanation many times to individuals. However, the danger exists that the teacher also can more easily yield to the temptation of lecturing more and more until soon the class has evolved back to the teacher-centered, large-group model.

Regardless of how the time is packaged, either into short modules or into longer periods, if the learning center design is the organizational basis for the department, then students will still have access to several teachers, to other learners, and to many resources, although they will be somewhat restricted in their access to time. If, however, the only available option is the one teacher–one classroom design, then students are further limited to interactions with a single teacher and a smaller number of fellow students. Even with this alternative, however, if mathematics classrooms are in reasonable proximity to each other, provisions can probably be made that would permit the movement of students back and forth among the several classrooms.

Faculties also can discover ways to make time more flexible, even within a fixed schedule. One approach is to schedule two or more classes back to back. For example, a group of sophomores might be scheduled for biology, geometry, and study hall during the first three periods of the day, with subsets of the students receiving schedules that are permutations of each other (see fig. 2.6). The biology and geometry teachers can then plan

| | Sophomore Students | | |
Period	Section I	Section II	Section III
I	Biology	Geometry	Study Hall
II	Study Hall	Biology	Geometry
III	Geometry	Study Hall	Biology

Fig. 2.6

that block of time according to need. One day all students might first see a biology film and then go to the mathematics center after the film. Another time one section might spend the entire morning in biology lab with no mathematics at all while the rest have all morning for mathematics and study hall. At still other times the activity might involve an interdisciplinary unit on the applications of mathematics in genetics. Figure 2.7 represents a hypothetical schedule in which only Wednesday is "regular." As an alternative, teachers might alter the schedule only one or two days a week and follow the fixed pattern the rest of the time. In this way students can extend somewhat their interactions both with other persons and with other areas of knowledge.

But scheduling is not the only factor to be considered. The curriculum, too, must be taken into account. One possibility is that regardless of the schedule, students could still gain access to the curriculum in very individual ways. Thus, those sophomores who arrive during first period might not all be studying geometry. Each could enroll for whichever of the short courses he or she elected—algebra, trigonometry, computer mathematics,

Period	I	II	III
MONDAY Sec. 1 / Sec. 3	Biology Film (all)		Math Center (all)
TUESDAY Sec. 1	Biology Lab		
Sec. 2	Math		Study Hall
Sec. 3	Study Hall		Math
WEDNESDAY Sec. 1	Biology	Study Hall	Math
Sec. 2	Math	Biology	Study Hall
Sec. 3	Study Hall	Math	Biology
THURSDAY Sec. 1	Math		Study Hall
Sec. 2	Study Hall		Math
Sec. 3	Biology Lab		
FRIDAY Sec. 1	Math	Study Hall	
Sec. 2	Math	Biology	
Sec. 3	Math		Biology

Fig. 2.7. Adapted schedule

or whatever. Then the fact that the entire group is composed of sophomores is purely accidental, and, in fact, students could be scheduled for that mathematics class regardless of their grade level or background. At reporting time, students receive the appropriate credits in whatever course or courses they have completed.

The question of grade reporting also is a consideration for teachers in traditional schools. Under the criterion-referenced system, a student may not have achieved mastery in a particular unit at reporting time. To accommodate that student, it is recommended that provision be made for assigning an I (incomplete), along with an explanation to the parents of the meaning and acceptability of that evaluation. If the department cannot maintain a testing center in which the student can take tests on request, tests can be obtained instead from the teacher in the center. If this seems unmanageable, a fixed day of the week can be designated as testing day. In the latter event, some students might be forced to wait a few days for a test, but while they wait they might be given the opportunity to proceed conditionally into the next unit.

These are only some of the adjustments that can be made to adapt the model to a school having a traditional type of scheduling. Whereas the open-school situation might facilitate some elements of the system, it is a serious fallacy to assume that the open arrangement is a necessary condition for its adoption or that without it none of the components of the model could be implemented.

Conclusion

A model for an alternative approach to secondary school mathematics through departmental reorganization has been outlined in this essay. An attempt has been made to identify both strengths and weaknesses of the program and to suggest approaches for implementing the system in schools having a variety of organizational patterns. Not only is the model one that can be adapted to different types of schools, it also can be implemented in schools of any size. Furthermore, it can produce significant changes without radical upheaval because it can—in fact, should—be implemented gradually and manageably. At the same time, it provides a true alternative to the study of secondary school mathematics—an alternative that can be exciting for both teacher and student.

3

Organizing for Individualization: A Classroom Model

Janet Barnard

Carol Dodd Thornton

*S*tudents differ in their attitudes toward learning mathematics, in their backgrounds and abilities, and in their interests in different mathematical topics. Many classroom teachers, sensitive to these differences, wish to develop a mathematics program that takes greater account of their students as individuals. A number of options for individualizing mathematics instruction, ranging from packaged commercial programs to self-made schemes, are open to these teachers. Budget restrictions or other administrative problems often prohibit full-scale adoption of commercial programs, and in so many schools teachers are left to their own resources (and limited finances) to design a mathematics program that fits the varied needs of their students.

It is to this group of teachers that this essay is directed. The model for mathematics instruction presented here can enable teachers to individualize inexpensively. It is flexible and can be implemented in a variety of teaching and learning situations, from the self-contained to the open classroom. The basic plan suggested by the model has been successfully adapted to a number of grade and ability levels, grade 4 through high school.

43

The Model: Individualizing Inexpensively

The major components of the general model for individualizing inexpensively are charted in figure 3.1. Whatever mathematics text has been adopted forms the basis for instruction, thus eliminating the need to purchase a more expensive program and materials. Students stay together by unit (chapter or topic) but are offered a variety of options within each unit. A diagnostic pretest, keyed to the fundamental concepts and skills of a unit, provides the basis for the subsequent assignment of learning tasks within the unit. Before the activities of an individual unit are started, however, the teacher introduces the principal concepts of the unit. Assignment sheets, which list learning experiences for the development of basic content as well as "second chance" and enrichment options, are then used to guide each pupil's study of the unit-related ideas. Project activities are interspersed throughout a unit's study plan to arouse interest and keep motivation high. A plan for the evaluation of student learning, based on diagnostic teaching and testing, is encouraged.

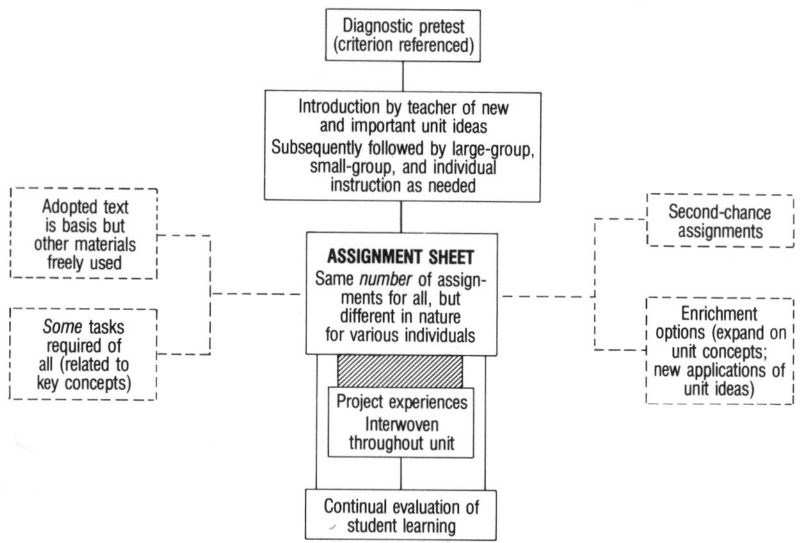

Fig. 3.1. General model for individualizing inexpensively

The model suggests an approach to mathematics instruction that allows for small- and large-group instruction, for peer interaction and independent study, and for observing, counseling, and working with individuals. Although competencies in mathematical understandings and skills are emphasized, attention is also focused on developing positive attitudes toward learning mathematics.

The model is designed to facilitate instruction for any mathematics unit. In the subsequent discussion a unit on number theory will be used as an example to show how the teacher's role relates to the major components of the model. This role essentially involves planning pupil activities and then directing these learning experiences throughout the unit. The progress of two students will further illustrate the actual implementation of the instructional model.

The teacher's role

Planning unit activities. In preparing a unit on number theory, for example, the teacher first would need to identify and list on a planning sheet the principal concepts and skills of that unit: divisibility; remainders; factors; common factors and greatest common factors; odd, even, prime, and composite numbers; multiples; and common multiples and least common multiples. The teacher next develops a *pretest* composed of a set of questions for each concept. When pretest questions are drawn from the adopted text series, an answer sheet such as that illustrated in figure 3.2 can be prepared (a sample question for each section has been given). Alternatively, pretest questions can be composed by the teacher or drawn from other inexpensive sources. It is suggested that a competency standard be established for each set of questions on the pretest. For example, out of five questions related to divisibility, the teacher might specify that a student must answer three or more correctly in order to demonstrate the desired proficiency. Using competency standards makes it possible to identify those students who can omit certain sections of a unit's work and thus have time for more challenging sections or applications of the unit ideas. A basic assumption underlying the design of the model is that students should work on mathematics content that matches their abilities and interests; one student's path through a unit will differ from another's.

From the flowchart in figure 3.3, it is seen that after formulating the pretest, the teacher would list on the planning sheet learning activities for each concept or skill to be developed. These activities can take the form of text assignments; worksheets and activity sheets; and laboratory experiments, puzzlers, and game assignments. They can be drawn from the adopted text series and from inexpensive workbooks and ditto sets to which the teacher has access. Commercial sources that grant duplicating rights are also helpful. Although it is nice to have a variety of resource materials on hand, one or two good ones are sufficient. Professional journals such as the *Arithmetic Teacher* and the *Mathematics Teacher* are a ready source of ideas. Some materials and aids can be purchased

PRETEST for Unit 6:   Name ___________________
Number Theory   Room _______ Date _____________

Use p. 61 of text:

+3 (1) Test 34: Divisibility
(Is 36 divisible by 9?)

     A. _______   B. _______
     C. _______   D. _______
     E. _______

+4 (2) Test 35: Remainders
$(25 = 6 \times 4 + \square.)$

     B. _______________
     C. _______________
     D. _______________
     H. _______________

+4 (3) Test 36: Factors
(List the factors of 12.)

     A. _______________
     C. _______________
     D. _______________
     F. _______________
     H. _______________

+3 (4) Test 37: Common Factors,
Greatest Common Factors

(Give common factors and
the greatest common
factor for 4 and 8.)

     D. _______________
     E. _______________
     F. _______________
     G. _______________

+6 (5) Test 38: Odd, Even;
Prime, Composite

(Identify 9 as even or odd;
prime or composite.)

     A. _______________
     B. _______________
     E. _______________
     G. _______________
     I. _______________
     K. _______________
     O. _______________

+4 (6) Test 39: Multiples
(List the first six multiples
of 7.)

     A. _______________
     C. _______________
     G. _______________
     J. _______________
     K. _______________

+3 (7) Test 40: Common
Multiples, Least Common
Multiples

(Give three common
multiples and the least
common multiple for 2
and 3.)

     A. _______________
     B. _______________
     D. _______________
     G. _______________

Fig. 3.2. Sample pretest answer sheet

from school-supply companies; even less expensive are materials and books from toy and grocery counters. Sets of activity cards or manipulative aids can be made rather than purchased.

In preparing for the unit on number theory, one teacher made a set of multiple sticks from a set of nine tongue depressors. On each stick nine congruent sections were marked off. The first nine (nonzero) multiples of 1 were written on the first stick, multiples of 2 on the second stick, and

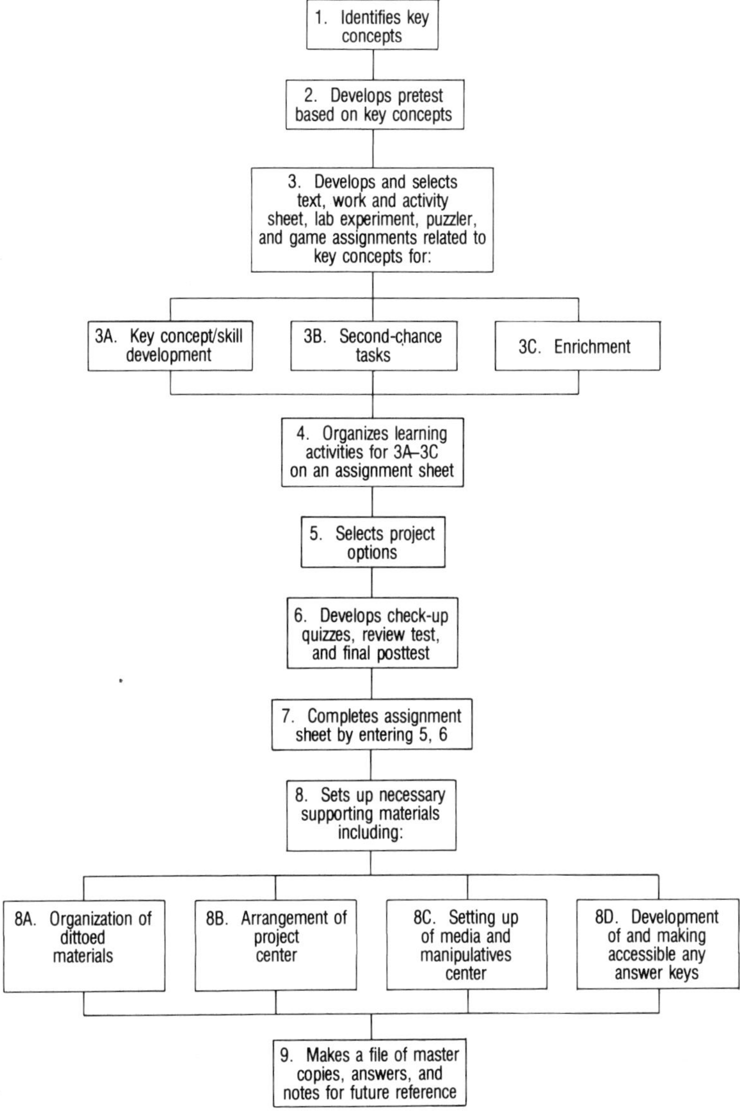

Fig. 3.3. Teacher planning stage

so on. The sticks were designed to help develop concepts and skills for finding common multiples and least common multiples. To find the least common multiple of 6 and 4, for example, the "4" and "6" sticks were aligned, one above the other, as shown in figure 3.4. The three multiples 12, 24, and 36 appear on both sticks; so 12 is the *least common* multiple of the two numbers.

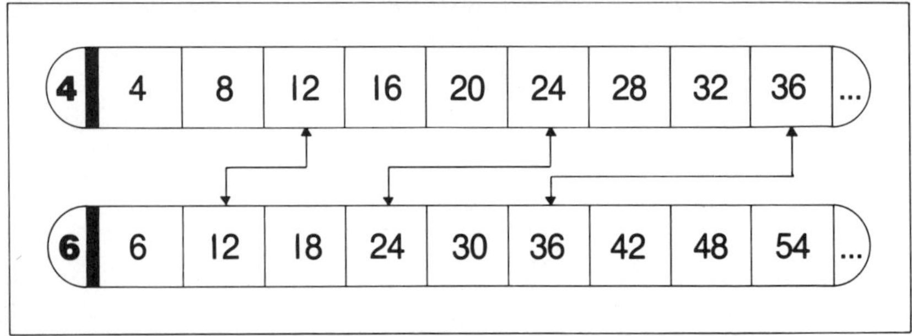

Fig. 3.4. Finding the LCM of 4 and 6 with multiple sticks

The learning activities the teacher has gathered and listed on the
planning sheet should now be grouped to coincide with anticipated student
needs. The basic need, of course, is for the development of the key concepts
and skills within the unit. Some activities to meet this need are developmental. Others are selected to follow up instruction by the teacher and
include discovery and pattern-finding experiences as well as practice and
review tasks. For example, a simple match-and-clip activity, illustrated in
figure 3.5, was used by one teacher to reinforce the more important ideas

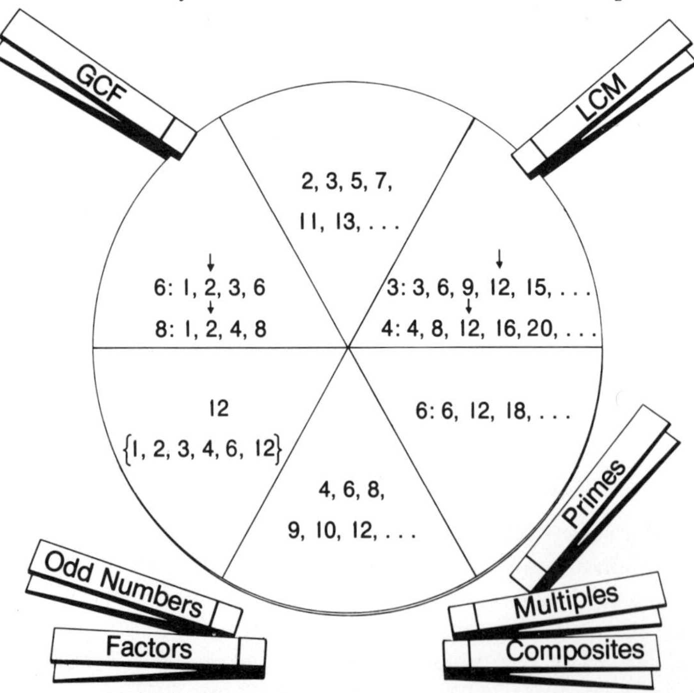

Fig. 3.5. Clothespin match-and-clip activity

in the unit. A second need is to provide for reinstruction. The teacher should be particularly mindful of those students who will not master a concept or skill the first time around. For these students alternative activities, some of which should use an approach different from that of the initial teaching, should be provided to give them another chance. Enrichment opportunities, for both faster and slower learners, comprise a third need. As students subsequently participate in unit activities, the teacher will accommodate their needs and interests by directing them to tasks from one of these three groups.

Within each of the three groups of activities the teacher selects the most appropriate experiences for developing the concepts or skills and arranges them sequentially from easiest to most difficult. An **assignment sheet** is then prepared by placing each group of learning experiences in a separate section on the sheet (fig. 3.6). *Section A* contains tasks for developing basic concepts and skills; *section B* lists the second-chance assignments for each concept of section A; and *section C* suggests enrichment options for the unit. Sections A and B are aligned so that second-chance opportunities corresponding to each learning activity of section A are located directly opposite that activity on the assignment sheet. For example, "workbook, p. 48," listed in section B, provides a second chance to master ideas and skills similar to those developed in "hardback, pp. 44–45."

The teacher completes the assignment sheet by indicating times for tests, quizzes, and project options and by putting a checkmark beside those assignments all students should do, regardless of how they performed on the pretest. All quizzes and tests are checked, as well as other selected activities that reinforce important concepts in the unit. The time period for the unit and the minimum number of assignments to be done by all students are also entered on the assignment sheet. This minimum number is determined by counting the total number of learning activities required to develop the basic concepts and skills of the unit, including tests and quizzes.

In this unit on number theory, a minimum of twenty learning activities is necessary to develop adequately the basic ideas of the unit. These activities are listed in section A of the assignment sheet. Even those students who demonstrate proficiency on certain portions of the unit must still complete at least twenty assignments, although they may include enrichment activities from section C as well. Students who need to spend time on second-chance activities will draw most of their twenty required assignments from sections A and B. For these students some of the more difficult work at the end of the unit may need to be omitted.

In determining the time limit for a unit, the teacher considers the number and length of the assignments. In this unit it was decided that most students should be able to complete at least two assignments each class period. Hence two weeks, 5 February to 16 February, were allocated for

Name _______________________

Room _______________________

Unit 6, Number Theory
Feb. 5–Feb. 16

ASSIGNMENT SHEET
(20 Assignments Required)

Column A

(A) ✓ Pretest: hardback, p. 61
_____ Puzzler
(1) _____ Hardback, pp. 44–45. Do A–Q (Cuis. Rds)
(1) _____ *Aftermath 2*, pp. 65–66
(1) _____ *Mathimagination Book C*, p. C36 Activity Sheet

Project
(2) _____ Workbook, p. 48. Do part 1
(2) _____ Play remainders game
(3) _____ *Aftermath 1*, p. 65, Prime Time Factor Activity

Project
(3) _____ Workbook, p. 49. Do part 1
(4) _____ *Mathimagination Book C*, p. C30, Line Up
(4) ✓ Workbook, p. 49. Part 2 (Use Cuis. Rds)

Project
(5) _____ Hundred Board Activity
(5) _____ *Aftermath 1*, pp. 13–14
(5) ✓ *Aftermath 4*, p. 24, Prime Pattern Activity

Project
(6) _____ Hardback, p. 54. Do A–U
(6) _____ Workbook, p. 50. Do part 2
(7) _____ *Aftermath 4*, p. 44, What Do We Have in Common Worksheet (Use multiple sticks)

Project
(7) _____ Workbook, p. 52. Do part 1
✓ Checkup Quiz— Workbook, p. 53
✓ Review test—Hardback, p. 58

Project

Column B

(B) _____ Workbook, p. 48. Do A–J
_____ *Aftermath 1*, pp. 39–41 Divisibility Activity

_____ Hardback, p. 48. Do A–M
_____ Factor Game
_____ Multiple-Factor Bingo

_____ Hardback, p. 49. Do A–L
_____ Hardback, p. 51. Do A–O
_____ *Mathimagination Book C*, p. C29

_____ Workbook, p. 50, part 1
_____ *Math 1st Dimension:* Circle Go Round Game, pp. 13–14
_____ Hardback, p. 52. Do A–M

_____ Multiple Sticks Activity
_____ *Aftermath 3*, pp. 18–19
_____ Multiple-Factor Bingo

_____ Workbook, p. 51. Do part 1
_____ *Mathimagination Book C*, p. C39

Column C

(C) _____ Workbook, p. 54, part 4
_____ Workbook, p. 55 (any 10)
_____ Hardback, pp. 60–61

Project
_____ Prime drag game
_____ Number Theory Puzzle Board
_____ Circle Vocabulary Match activity

Project
_____ *Cloudburst of Math Lab Experiments* (Any 3 from pp. 21–29)
_____ *Math Experiments with the Number Tablets* (Ewbank), any 3 on number theory
_____ *Eureka*, p. 33, Sieve of Eratosthenes

Project
_____ *Math 1st Dimension*, p. 35
_____ *Mathematics a Human Endeavor*, pp. 44–45, Set 1; p. 58, problem 5
_____ *Mathematics a Human Endeavor*, pp. 6–7, III

Project
_____ Read *Giant Golden Book of Mathematics*, pp. 18–19, "The Shapes of Numbers"
_____ Read Booklet *Prime Numbers* by H. Larsen
_____ *Mathimagination Book C*, p. C25

Project
_____ *Mathimagination Book C*, p. C35
_____ *Mathimagination Book C*, p. C45
_____ *Aftermath 1*, p. 50, "Multiple Madness"

Project

Column D

Project
_____ *Aftermath 1*, pp. 85–86 (Pascal)
_____ Fibonacci Sequence Activity
_____ *All Sorts of Numbers* (Bell), any 3

Project
_____ *Aftermath 2*, p. 33, Perfect Nbrs.
_____ *Aftermath 2*, p. 48, Patterns and Sequences
_____ *Working with Leftovers* (McCutcheon) Residues (any 3 activities)

Project
_____ *Aftermath 2*, p. 53, Prime Factor Tents
_____ *Aftermath 3*, p. 12, Divisibility
_____ *Patterns and Puzzles in Modern Math* (Graflund), Goldbach's Conjecture, p. 15

Project
_____ *Aftermath 3*, p. 35, Factors
_____ *Aftermath 3*, p. 69, Number Magic
_____ *Aftermath 4*, p. 81, Common multiples

Project
_____ *NCTM 35th Yrbk*, Triangular & Square Numbers, p. 490
_____ *Aftermath 4*, p. 87, Primes
_____ *Aftermath 4*, p. 90, Twin Primes

Project

completing the twenty required activities. The time limit may need to be extended if the teacher has misjudged the amount of time needed to complete unit work, but if only a few children lag behind, the teacher would probably move on to a new unit, satisfied that even these students had mastered the most basic ideas.

Having finished the assignment sheet, the teacher next prepares **project options.** In practice, these can be highly motivating, can cultivate student interests, and can serve to extend, apply, reinforce, and review the important concepts and skills of the unit. Both project and enrichment activities would be drawn from similar sources, but projects are selected particularly to provide a motivational change of pace for slow as well as quick learners. In selecting the projects, the teacher would consider including activity cards, puzzlers, laboratory experiments, instructional games, or studies of selected mathematical topics of special interest. Projects may well take on a multimedia flavor through the use of printed materials (including appropriate library references), audiovisual resources, manipulatives, and electronic calculators or computer terminals, if available. Often a teacher will discover that materials from libraries or other instructional and media centers fit a unit's theme and can be used without cost.

In the unit being used for illustration, students are given the option to select a project and work on it for fifteen minutes after the completion of every three assignment tasks. An alternative might be to set aside a project day from time to time.

The next step in the planning process is to select the means for evaluating pupil growth in the unit's concepts and skills. Aware of the scope and sequence of the unit, the teacher can prepare checkup quizzes, a review test, and a posttest. The purpose of the checkup quizzes and review test is to help the teacher determine any needs for further developmental work. The posttest, parallel in form to the pretest, is the final means of assessing the students' mastery of the key concepts and skills of the unit. The teacher might also want to use other informal measures, such as observations and conferences with students. Establishing a filing system for each individual's completed work and project activities would further help the teacher in making a final evaluation.

Students should be involved in a day-to-day evaluation of their learning. To make them aware of their progress, the teacher can plan several ways to provide immediate feedback. A system can be set up whereby students grade their own pretests and daily work and do self-checking activities. Grading quizzes and the review test *with* each student as these are completed is another means of cuing individuals to their progress. During such sessions the teacher may choose to give partial credit for corrected quiz and test items.

To complete the planning of the unit, the teacher next procures and organizes the necessary supporting materials. Dittoed materials can be placed on shelves of a homemade wooden cabinet or filed in a spare file drawer or box. It is helpful to arrange project activities and their accompanying materials in a Project Center, located in one corner of the room. Materials can be inexpensively stored and organized in zip-lock bags, shoeboxes, or large mailing envelopes. Media and manipulative materials should be made readily available by storing them in another section of the room.

One of the teacher's final tasks in preparing a unit is to develop answer keys for the student activities and tests. Students should have access to all keys except those for quizzes and tests. Answers for the pretest could be written on a transparency so that all students can check their work at the same time. All master copies, answers, and teacher notes should be filed for future reference.

Implementing unit activities. The teacher's first responsibility in implementing a unit of work is to administer the pretest to students. Puzzlers can be provided to those students who finish before others. The teacher then supervises the checking of the test by students. (In checking their own work, students are made more aware of their specific needs.)

Now that attention has been focused on the concepts and skills related to the unit and individual needs have been identified, the teacher provides a preview of the unit to introduce the more important concepts and skills to the entire class. This **initial introduction** sets the stage for the subsequent development of ideas for those students who did not demonstrate com-

petency in particular areas. For the more able students, this session helps to clarify ideas and procedures associated with the more basic concepts and skills and gives them a foundation for beginning work on the more difficult sections of the unit.

After this introduction, the teacher supervises the students as each plots a plan of work for the unit on the assignment sheet. Then, as students follow their individual paths through the unit, the teacher's role becomes multifaceted. The teacher is continually involved in diagnostic teaching through answering individual questions, providing small-group instruction as needed, and occasionally working with the entire group to develop, review, or clarify an important idea. Short, whole-class review quizzes, boardwork, or quick oral reviews—given at the beginning of a period— help in diagnosing problems, in tying together parts of the unit, and in reaching the entire group in an area of special need. The teacher observes and interviews students at work, grades checkup quizzes and review tests,

and redirects student learning when necessary. At the end of the time allotted for work on the unit, the final posttest is administered and graded.

The student's role

In practice, the student's role in the model is one of active involvement in the mathematics to be learned. This involvement entails working independently as well as cooperatively with others and interacting with the teacher as needed. The student is encouraged to use materials and other learning aids and is free to make choices and decisions related to the learning (e.g., the selection of project topics and enrichment activities). Working in a relaxed classroom atmosphere and being free to move about as needed, students create a certain amount of noise. But since the students are organized and have purposeful direction toward definite goals and since each is encouraged to respect the right of others to learn, this atmosphere should be busy yet controlled, so that a positive learning environment is assured.

Bryan and Sherri's class is about to begin a new unit on number theory. After being reminded by their teacher that the purpose of the unit pretest is to find their areas of strength and weakness and that the pretest will not be used for grading purposes, each answers as many questions as possible. Sherri, feeling quite familiar with the ideas on the test, works slowly and carefully, and completes all but one or two of the pretest questions. Bryan, however, is stumped by most questions. He guesses at a few but leaves many blanks. Quickly finished with what he can do, he looks over the puzzlers that have been provided. One is a color-coded puzzle that reviews basic computation. The other presents a riddle that can be solved using simple ideas from the unit on number theory. Bryan chooses to do the puzzle.

All pupils in the class grade their own pretest from the transparency projected on a screen. Bryan has no trouble in quickly grading his test. Sherri questions whether some of her answers are equivalent to those shown—her grading takes a little longer. During the teacher-directed introduction to the unit which follows, Sherri resolves a problem that had confused her since taking the pretest. For Bryan, most of the ideas are new. He copies a diagram that seems to be helpful.

Along with others in the class, Sherri and Bryan next learn how to fill out their assignment sheets. First they notice the seven concept sections of the pretest (fig. 3.2) and the circled numbers corresponding to each. These same numbers appear in sequence on section A of the assignment sheet (fig. 3.7), and they match in theme the related section on the pretest. Section 1 of the pretest, for example, deals with divisibility. On the assignment sheet three learning tasks deal with divisibility and are also labeled 1 (all section numbers have been circled on both the pretest and the assignment sheet). Bryan looks at each section of his pretest to determine whether he has met the competency standard, the minimum number of correct answers needed to bypass assignments on a given aspect of the unit topic. The competency standard for section 1, divisibility, is three correct answers (denoted by the +3 in the margin). Bryan fails to demonstrate competency in this section and in sections 2, 4, 6, and 7 as well. Realizing that he needs to work in these areas, Bryan places checkmarks next to the assignments for those sections (see fig. 3.7).

Sherri's pretest results diagnose that she needs to concentrate her study in sections 4 and 7. These, then, are the assignment sections that she checks. In addition, certain selected assignments, quizzes, and tests have already been checked by the teacher for all students.

With their individual assignment tasks plotted, students begin independent work on the unit. Bryan gets the Cuisenaire rods that are called for in his first learning task. Sherri goes to a nearby file to get her first activity sheet. Others in the class may find that their first assignment is in the text. All students, on completing a task, use the answer keys to grade their own work.

Bryan understands the ideas involved in the first assignment and finds, after grading his work, that he did quite well. This task completed, he crosses it off on the assignment sheet, files the work in his folder (kept on a shelf in the room), and moves on to the second assignment. In grading this learning activity, he finds he missed more than three problems, and

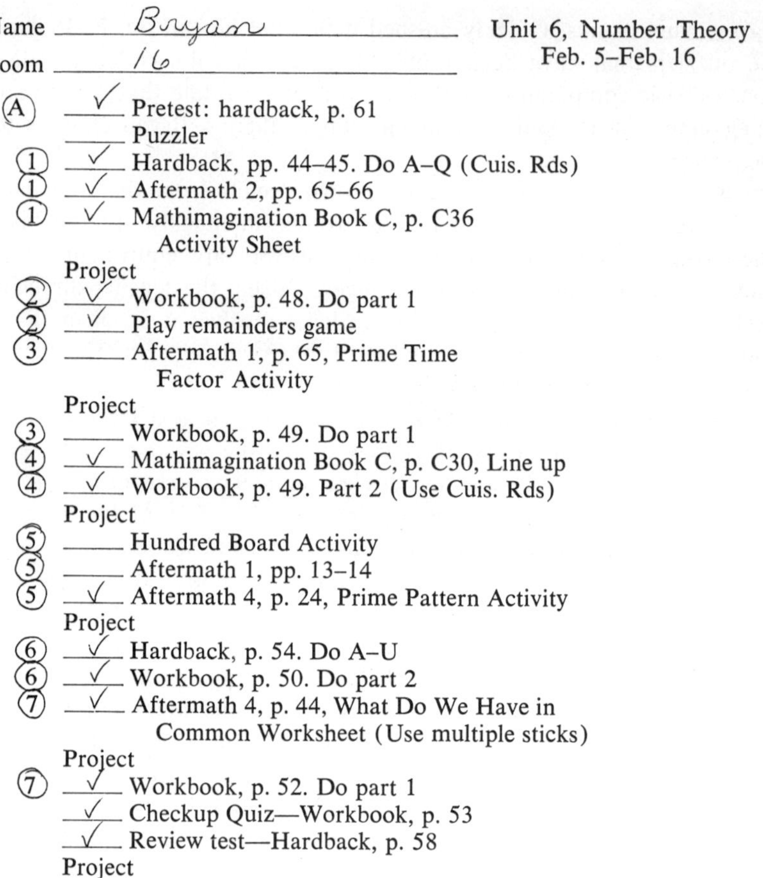

Fig. 3.7. Section A of Bryan's assignment sheet

so he goes to the teacher for additional help. With this extra instruction Bryan is able to correct the mistakes and do the second-chance assignment that the teacher substitutes for a harder learning activity at the end of the chapter. (Bryan's class had previously been instructed that they were to check with their teacher when missing more than three problems on any assignment.)

Sherri had a perfect score on her first assignment. When she graded her second assignment, she found two mistakes. She corrects these, crosses off the check on the assignment sheet, and goes on to the next activity. When this third assignment is completed, scored, and corrected, Sherri will be ready for a project option.

Giving a student a choice in selecting project activities of interest has proved, in practice, to be an important element in the development of posi-

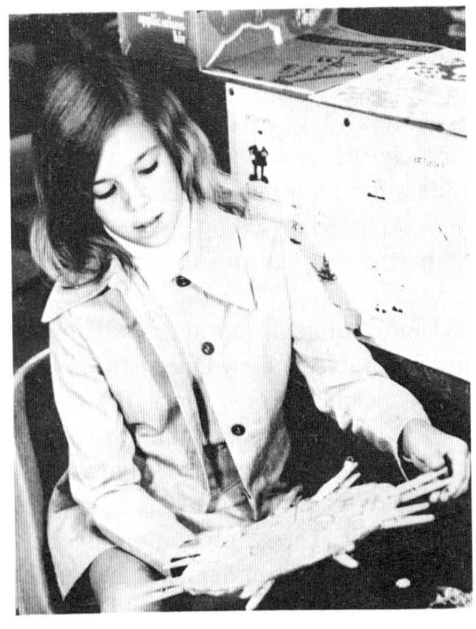

tive attitudes toward mathematics learning. Therefore student project options are considered essential. All students—whether fast or slow learners—are encouraged to take time out to enjoy mathematics in an area that is of interest to them personally.

Before going to the Project Center in the classroom, Sherri writes her name and the time on a designated section of the chalkboard. She then selects a project in which she is interested and begins to work on it for the fifteen-minute time period. She notices that two of her classmates are also in the center, working together on a project, but on this occasion Sherri prefers to work independently. At the end of her project time, Sherri replaces the materials she had been using and erases her name and the time from the board. Returning to her seat, Sherri makes a brief record of the activity in her project notebook. At a later time she will show the book to her teacher to discuss the types of activities she has been doing.

As work progresses throughout the unit, Bryan and Sherri become involved in a number of different learning situations. When Bryan discovers that he and two of his friends have similar problems, they meet in a small group with the teacher for special instruction. When Sherri has a specific question further on in the unit, she asks the teacher for clarification. At the beginning of one class period, all students are involved in a quick review of the basic ideas of the unit. Sometimes Sherri finds there is less distraction when she works alone in a quiet corner of the room. Bryan, however, seems to learn more when working with a friend on the same assignment. Two can often figure out what one cannot. Toward the end of the unit, Sherri turns for help to another student, who is able to answer her question. Bryan, preferring to spend more time mastering the work in section A of the assignment sheet, decides not to take his next project time. He continues working sequentially through the checked learning tasks of section A until he is ready for the review test.

As with the checkup quizzes, Bryan works independently to answer the questions of the review test and then asks his teacher to grade it. He watches his test being graded and works with his teacher on problems he missed. Bryan then returns to his seat to try a second-chance assignment he has been given. This assignment, suggested by the teacher, is a follow-up to the teacher's instruction to assure that Bryan really understands the ideas presented.

Sherri, having taken the review test several days earlier than Bryan and having had few problems with it, spends the rest of the time selecting enough enrichment tasks to meet at least the required number of unit assignments. Both Sherri and Bryan are expected to complete a minimum of twenty assignments, as determined by the teacher and noted at the top of the assignment sheet.

For Bryan, the teacher has replaced more difficult learning assignments with second-chance tasks that are more profitable to him. One such second-chance assignment involved the use of Cuisenaire rods. Since Bryan was having difficulty understanding the work on primes and composites, his teacher suggested the rod activity. Using the rods, Bryan could show that 1 and 7 are the only factors of 7, and hence 7 is prime; yet 8 is composite because it has other factors besides 1 and itself. The manipulation involved in the activity helped Bryan visualize the ideas and understand better the difference between prime and composite numbers.

Even though Sherri and Bryan both do a minimum of twenty learning tasks, they do not necessarily do the same amount of work. Assignments toward the end of the unit are more difficult, more detailed, and more thought-provoking. On the one hand, students like Sherri will often finish these and have time to work for extra credit from the remaining optional assignments. On the other hand, slower students like Bryan do shorter and more basic learning tasks on the premise that it is better to learn well the most essential concepts and skills of the unit than to half-learn everything.

On the last day devoted to the unit's work, Bryan and Sherri take the final posttest on the unit. This test is a final evaluation of the unit's concepts and skills and is therefore graded by the teacher outside of class time. Bryan and Sherri receive their grades and corrected test papers on the following class day.

Classroom Arrangement

Flexibility is perhaps the most important element in setting up the classroom for the effective implementation of the model. The layout of the learning area, the classroom furniture (shelves, tables, desks, files, etc.), the amount and kind of supporting learning materials to which the teacher has access, and the number of pupils involved will be major factors in determining an effective classroom arrangement. Within the learning area, furniture will need to be arranged and rearranged from time to time to accommodate different instructional settings. The teacher must provide for a project center, a media and materials center, a checking station, and a file for printed materials. Storage space for student folders, a shelf for reference materials, and some place in the room where the teacher can work individually with a student should also be provided. By being creative and flexible with materials and the classroom setting, teachers will find that they have at their disposal the means to individualize inexpensively.

Shelves, tables, or other room dividers can be used to separate the project center from the rest of the room. It is helpful to choose for media and materials a section that has an electrical outlet and shelf or desk space for the storage of student activities. Often it will be possible to section off this area as well. Large pieces of cardboad can be substituted for more expensive room dividers (cardboard cutting boards, purchased on sale at fabric or sewing centers, are excellent for this purpose and are easy to store). To cut down on room noise and to create a more relaxed atmosphere, carpeting, rugs, or toss pillows can be used in these areas. Many stores will donate rug samples or make them available to teachers at a nominal cost.

The checking station contains answer keys for student checking. A table, shelf, or desk can be used to house the keys as well as to provide the students with a place to check their own work. Answer keys should not be taken from the checking area.

Worksheets and activity sheets, along with any other dittoed material, can be stored on shelves, in a filing cabinet, or in a box located conveniently in the room. Homemade filing shelves have proved to be a very efficient means of storage. Such a cabinet can also double as a room divider. Storing students' assignment sheets, completed work, and tests in a folder in the classroom eliminates lost and forgotten assignments and gives the teacher ready access to a file of a student's work. This arrangement is particularly useful for parent conferences and in assigning work for absent students.

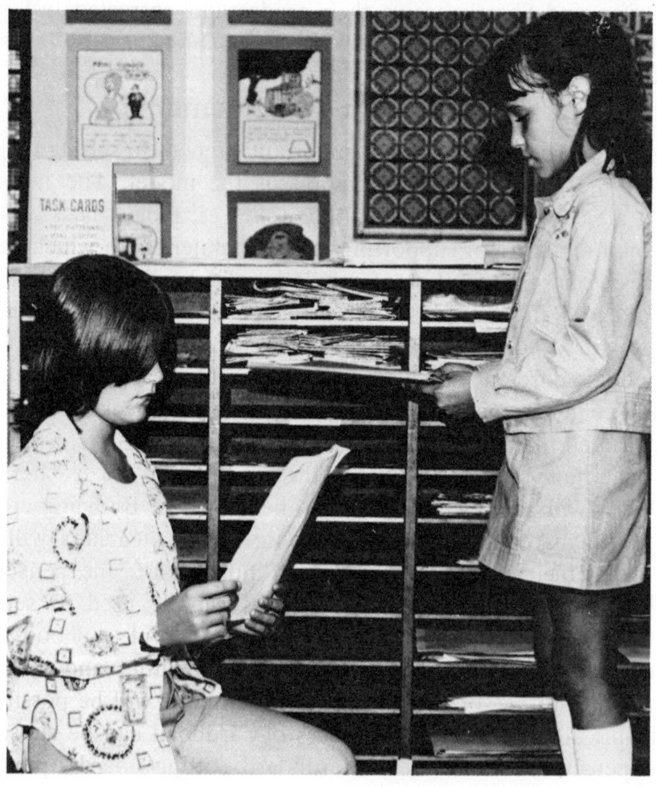

A reference shelf should be located somewhere in the room. Necessary references will vary from unit to unit, but an atlas, an almanac, a dictionary, and an encyclopedia are always useful. Mathematics library books may be kept in this area as well.

At times a teacher may wish to set up other learning stations to correlate certain ideas with a unit theme, to match student interests, or to reinforce a particular idea or skill. These stations can be partitioned from the rest of the room by dividers or, if space is limited, set up inexpensively on bulletin boards.

Adapting the Model

A teacher sometimes has little control over the general instructional organization within the school building, since this is often an administrative decision. From school to school and even within a building, instruction can asume a variety of forms, from team teaching, departmentalization, and the self-contained classroom to pod arrangements, multiple-grade or multiage classrooms, and more open situations. The flexibility of the model presented in this essay allows it to be adapted to a number of different instructional settings.

Although in practice *team teaching* is used in a variety of ways, the idea is one of teacher cooperation in planning and teaching a unit. Since the model approaches mathematics through unit themes, very little adaptation is required for team teaching. In planning and implementing a unit, teachers can either work together in developing ideas or take turns serving as unit leaders.

In most settings, including *departmentalized situations,* unit planning makes it possible to provide a broader scope and a more continuous sequence of instruction than that which is possible in day-by-day lesson planning. The departmental mathematics teacher, being responsible for a particular subject area, is able to concentrate full time on mathematics planning and teaching. Arranging to stay in one room throughout the day further facilitates setting up the program outlined in the model.

In the *self-contained classroom,* the teacher can set aside one section of the room for mathematics. The teacher who is with the same students all day is in a position to keep abreast of special interests and needs and can relate mathematics learning to a student's strengths in other areas as well. This teacher, being responsible for lessons in all subject areas, can work ahead in mathematics through unit planning.

Large *pod systems* usually involve flexible and cross-grade grouping. Unit planning in mathematics, as suggested in the model outline, is well suited to this type of structure, since it is possible to regroup from unit to unit on the basis of a student's strengths, needs, and interests.

The teacher in a *multiple-grade classroom* has two options: (1) developing separate unit plans for each grade or (2) working with the entire group to develop ideas within a particular content area. In choosing the latter option, the teacher draws the concepts and skills of each grade level together into one unit plan. For example, a teacher with a combination seventh- and eighth-grade class may wish to develop with the entire group a unit on integers. The teacher would select ideas from both texts in developing the pretest and in designing the assignment sheet. This provides for both lower-level remedial work and higher-level extension activities that match individual needs.

In the *open classroom,* the model provides a format for the organization of mathematics learning that allows for a sequential development of concepts and skills and respects individual pacing. From the standpoint of the teacher, instruction is more manageable in this format because students are together, at least by unit topic. At the same time, the freedom of students to make decisions and to choose activities related to their learning is still protected.

Within the framework of each of these instructional organizations, a variety of teaching methods or procedures can be incorporated. Depending on a teacher's general philosophy and preferences in teaching, assignments can be designed to stress a particular approach or combination of approaches. By emphasizing the use of guided discovery, laboratory learning, activity packets, learning centers, lectures, drill-and-practice sessions, or discussions, teachers can tailor instruction to reflect their own ideas and style of teaching. Each teacher will want to take into account the learning styles of individual students as well. The model is flexible enough to allow varying degrees of structure. Take, for example, the child who needs more teacher interaction and cannot function independently. The teacher can deal individually with such a child by assigning appropriate learning activities on a daily basis and by checking with the student at the completion of each task. Though diagnostic teaching is built into the program, a teacher may want to emphasize this technique to a greater extent with special students. More rigid competency-based standards can also be established in planning and implementing the program.

Concluding Remarks

In practice, students involved in a learning situation such as that described in this essay have developed and demonstrated positive attitudes toward mathematics learning. The reason for this may lie in the fact that an effort has been made to make mathematics learning interesting, challenging, and fun—within the bounds of reasonably attained goals (and, from the teacher's standpoint, within the bounds of limited budget allowances). The model was designed with the realization that having an active part in the selection of many of their learning experiences and being free to interact in learning situations with others are key factors that enhance students' learning atmosphere and contribute to attitude development. Moreover, realistic, short-term goals give pupils a reasonable chance of demonstrating mastery. Being continually involved in the evaluation of their own work, students are aware of their successes and can sense their progress toward established goals.

4

Organizing Independent Learning Units

Joseph Abruscato

Clinton A. Erb

*T*eachers can use many strategies to help children develop the ability to think and act independently. Although one specific way of providing such experiences will be discussed here, we do not suggest this method as a total approach to mathematics education. Indeed, recent research indicates that an independent learning format, when used as the exclusive learning mode, may not be superior to more traditional approaches (Schoen 1976 [a] and [b]). Students need a variety of learning experiences; the Independent Learning Unit (ILU) represents just one component in the teaching repertoire.

In this essay the following questions will be discussed:

1. What are independent learning units, and how can they be developed?

2. How can the materials and the classroom be organized so that independent learning units can be used successfully?

Designing Independent Learning Units

The Independent Learning Unit (ILU) is a self-contained set of sequenced activities and materials that students work through with a minimum of assistance from the teacher. A complete ILU has three basic components (see fig. 4.1):

1. *An activity card deck:* a carefully developed set of directions to engage the student in specific activities
2. *A materials package:* all the materials, including games, worksheets, audiotapes, and so on, that are required for the given activities
3. *An assessment package:* the evaluation materials—a preassessment, which comes at the beginning of the unit, and a postassessment, which is used at the conclusion of the unit—needed to assess the change in the student's abilities that results from completing the unit

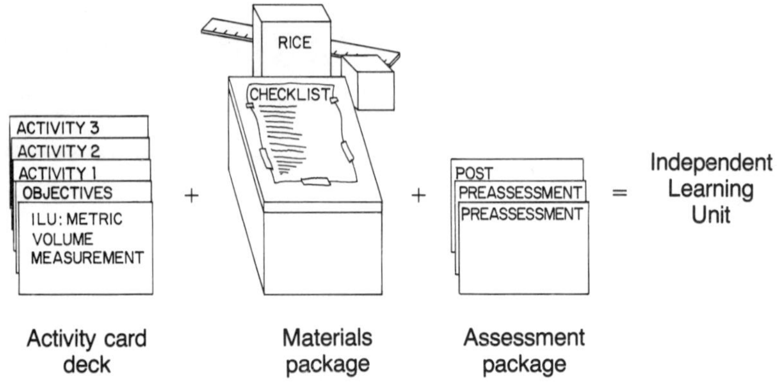

Fig. 4.1. Components of an ILU

The ILU is designed to assist the student in learning mathematical skills, concepts, or processes that can be acquired in a short period of time—perhaps four or five work sessions. Short units provide for flexibility within the total mathematics program, since students do not have to be committed to an extended amount of independent work. The short time span also enables students to maintain their motivation throughout the unit; the end of a particular ILU is near enough to the beginning to make the student feel that successful completion can be a reality. Finally, the teacher can obtain frequent evaluations of the student's progress from the assessment portion of the ILU.

Selecting a topic and establishing objectives

The needs of the student are of primary importance when selecting a topic for an ILU. Part of the student's experience in school should be of an independent nature so that self-confidence and the skill of working individually can be developed. The topic of an ILU should lend itself to this type of work and should be adaptable for presentation in the ILU format. Some topics may be too involved to be clearly presented in such a manner.

The ILU can complement the total mathematics curriculum and provide for increased flexibility in responding to diverse student needs. ILUs can be developed for remedial as well as enrichment purposes. They can be used as an alternative method of instruction in a topic that is an integral part of the curriculum. ILUs can also be used as a method of review as well as a way to introduce new concepts.

There are many ways of generating topics for an ILU. The class could be asked to develop a list of topics they would like to investigate on an independent basis. Diagnosis may indicate that certain topics need to be reinforced for certain students. A group of teachers who wish to share ideas could develop a list of topics together. Then each one could develop separate but coordinated units that all could use. A list of illustrative topics for independent learning units is shown in figure 4.2.

Primary Level

Prenumber concepts of sets	Place value
The commutative principle	The recognition of circles and squares
Squares of numbers	The concept of large and small
Counting money	Equality and equivalence of sets

Intermediate Level

Trichotomy law	Finding sums of unlike fractions
Areas	Multiplication of fractions
Plotting coordinates on graphs	Concept of symmetry
Number patterns	Measuring dry weights

Secondary Level

Basic conic sections	Introduction to vector geometry
Surface areas and volume	Introduction to topology
Perimeter	Complex numbers
Networks	Binomial expansion

Fig. 4.2. Sample topics for possible ILUs

Preparing the activity card deck

The activity card deck serves as the students' basic guide by providing directions for the activities. For those with poor reading skills, an audio-taped set of directions should accompany the cards. Using the activity card deck, the student works through a group of experiences that include the following:

1. *Introduction.* Through a conference with the teacher, the reading of an introductory paragraph, or some other method, the student is introduced to the unit and learns of its objectives.

2. *Preassessment.* Through the early activities of the unit, the student and the teacher determine the student's entry-level abilities with respect to the objectives of the unit. These activities ascertain if the student has already reached the objectives, which might make working through the unit unnecessary.

3. *Learning activities.* Through these activities—the heart of the unit —the student becomes involved in independent action. The specific activities are designed to apply such necessary learning requisites as active student involvement, logical sequence, repetition, and positive reinforcement through self-correction and the knowledge of results.

4. *Postassessment.* Through a special activity, the level of success in the unit is determined.

5. *Conference.* Although the teacher may have communicated with the student during the unit, time is set aside at the end so that together they can plan the next appropriate learning experience.

Each of these components will be considered briefly and illustrated with a component from an ILU that was developed to introduce metric volume measurement to sixth graders. This particular unit was developed as an alternative to presenting the material in a large-group situation.

The activity card deck is based on the objectives for the unit. To provide a truly independent learning experience, the teacher must be sure to express objectives clearly. It is important that the student fully understand the purposes and direction of the unit and exactly what is expected. This is facilitated when the objectives of the unit are expressed in terms of student performance. Three major types of objectives are cognitive, affective, and psychomotor; an ILU may be designed to help students attain one type of objective or a combination of them.

Introduction and objectives card

This card is designed to introduce the ILU and its specific objectives to the student. (See fig. 4.3 for the introduction and objectives card from a sample ILU, "Metric Volume Measurement.")

Preassessment card

The purpose of preassessment is to determine whether the student has

---

Introduction and Objectives Card

This unit introduces you to the metric measures of volume. In order to complete this unit, you should be familiar with the metric units of length, including their symbols, and be able to measure using a decimeter ruler. You should also know the basic metric subdivisions and the following prefixes: milli-, centi-, deci-, and kilo-.

Objectives:

After completing this unit, you will be able to—

1. explain the relation between the metric units of length and volume, including the relation between cubic centimeters and milliliters;

2. compare the relative sizes of a milliliter (cubic centimeter), liter (cubic decimeter), and kiloliter (cubic meter) and place them in order from smallest to largest;

3. approximate the volume of a given container in metric units using rice and a box having a volume of one liter.

Fig. 4.3

already mastered the objectives of the unit, and so it is extremely important that the preassessment relate directly to the unit objectives. If the objectives have been achieved, the student could then be referred to other ILUs or to other learning experiences that would be more appropriate.

Preassessment activities can be written, performance-oriented, or oral in format. (See fig. 4.4 for an example of the types of questions and activities that could be designed. The actual preassessment would include

---

Preassessment Card: Written

Don't worry, this is NOT a test! Do the best you can.

In the folder marked "ILU: Volume Measurement" on my desk, you will find a ditto labeled "Preassessment Worksheet." Answer the questions on that paper and put the completed sheet in your ILU folder. Continue with the performance and oral parts of the preassessment. When you have completed all three parts—written, performance, and oral—bring your folder to me so we can discuss whether you should continue in this ILU.

Fig. 4.4a

Preassessment Worksheet

Remember, this is NOT a test! Do the best you can. When you have finished this sheet, place it in your ILU folder.

Circle the choice that is the best answer.

1. The volume of a gallon is about
   a) 3.75 ml
   b) 3.75 l
   c) 375 ml
   d) 375 l
2. The best unit for measuring the volume of an aquarium is
   a) cubic dekameters
   b) milliliters
   c) liters
   d) cubic meters
3. Arrange the following volumes from smallest to largest:
   cubic meter, cubic centimeter, liter
4. What is the volume, in liters or milliliters, of a box that is 25 cm long, 10 cm wide, and 8 cm high?

Fig. 4.4*b*

Preassessment Card: Performance and Oral

Performance

Using some rice and a box that has a volume of one liter, determine the volume of a Matchbox Car box. You will find the Matchbox Car box in the box labeled "ILU Materials." Hint: How many of these boxes does it take to fill the liter box? After you have your answer, bring your results to my desk.

Oral

Come to me and explain the relation between the metric units used for measuring length and those for measuring volume. What is the specific relation between a cubic centimeter and a milliliter?

Fig. 4.4*c*

Fig. 4.4. Preassessment cards and worksheet for an ILU entitled "Metric Volume Measurement"

a more comprehensive list of questions or tasks.) The format of the pre-assessment activities depends on the objectives. For some ILUs, all three types—written, performance-oriented, and oral—will be needed; for others, one or two will be sufficient. As teachers consider and develop the preassessment, they must refer to the ILU objectives so that the measures obtained are consistent with the expected outcomes.

Learning activity cards

The assumption underlying the development of the activities portion of an ILU is that students can progress through these experiences in a relatively self-directed manner. The teacher assists when the need arises and provides as much encouragement as necessary.

The activity cards direct students to the experiences that should enable them to attain the objectives of the ILU. Each activity card should provide the student with either a major activity or a group of smaller activities that could reasonably be carried out during one class period. The number of activity cards needed will depend on the unit and the scope of its objectives. (See fig. 4.5 for the third activity card of the sample ILU. In the first two activity cards for this unit, standard measures of volume and the relationship among them are explored using Cuisenaire materials.)

All activities used in the ILU should meet four basic criteria, which are drawn from contemporary beliefs about how students learn:

1. Students must be *active* participants in the learning experience.
2. Students learn basic skills more efficiently when experiences are *logically sequenced.*

Activity 3

In the last activity, you found that the standard unit of measurement for volume is the _____?_____. What relation does the cubic centimeter have to this unit?

Go to the folder labeled "ILU: Volume" and take out a sheet entitled "Activity 3." On this sheet list three (3) quantities that would be best measured in liters and three (3) that would be best measured in milliliters. Keep in mind the sizes of a liter and a milliliter.

Get the box that you made in Activity 2, which had a volume of one liter. Use this box to approximate the volumes of other containers. Get a quart milk carton from the box labeled "ILU Materials." Which do you think is larger, a liter or a quart? Fill the quart container up to the top with rice. Pour the rice into your liter box. Was your first guess correct? Although there is not much difference between a quart and a liter, the liter is larger.

(continued)

Activity 3 (cont.)

How many milliliters are contained in the drinking glass that is on my desk? First estimate your answer and write it down on your paper. Using your liter box and the rice, can you figure out the volume of the glass? Here are some hints that will help. How many glasses does it take to fill the box? How many milliliters does the box contain? Can you put these two facts together to get your answer? If you are still having trouble doing this, see me. Write your answer on your paper. How close was your estimate?

Here is one last question that you can figure out with your box and rice. Approximately how many milliliters of milk are contained in the carton that you drink for your snack? Write your answer on your paper and put the paper in your ILU folder.

Fig. 4.5. Sample activity card for an ILU on "Metric Volume Measurement"

3. Appropriate responses to the situations that are posed in the materials should be *immediately reinforced.*

4. Students should periodically *repeat* the experience in order to retain what they learn.

Active involvement. Students must be actively involved in the learning experience to maintain interest. Different media may be employed so that they can respond to a wide variety of stimuli. They can assemble models, perform experiments, or use appropriate manipulative activities. Making liter boxes and conducting experiments with rice are examples from the sample ILU which illustrate this point (see fig. 4.6).

Fig. 4.6

If commercial games are used, they may have to be modified to fit the purpose of the ILU. Both commercial games and those developed by the teacher must be directed to the appropriate level of student sophistication in order to be fully effective.

Sequencing. One of the most challenging aspects of developing an ILU is the proper sequencing of experiences. Often, this will need to be revised as the students' progress is observed. Appropriate sequencing will be easier as experience is gained in developing ILUs.

Reinforcement. To encourage progress, each daily lesson should include opportunities for students to be made aware of the correctness of their work. This can be accomplished by using answer keys or by supplying the correct responses during the course of the activity. Reinforcement can also be presented through conferences and through comments made to the student on the sheets that are handed in at the end of each activity.

Repetition. To help ensure that the student progresses steadily toward the unit's objectives, activities should provide for repetition in an interesting fashion. In the sample ILU, the student handles a square decimeter from the Cuisenaire materials in activity 1 and then builds a box of the same size in activity 2. In activity 3, the student is helped to determine the volume of a drinking glass in milliliters and then independently calculates the volume of milk consumed during a snack.

Postassessment card

After students have completed the activities, an evaluation should follow to determine if they have attained the objectives of the unit. This is accomplished through the use of written, performance-oriented, or oral postassessment measures. The postassessment card directs the student to the assessment activity. (See fig. 4.7 for the postassessment card of the sample ILU.) Since the student should be provided with experiences in postassessment that are identical or similar to those in the preassessment, both activities can be developed concurrently. The teacher should be certain the assessment relates directly to the purposes of the ILU.

Students completing independent learning units will show varying degrees of achievement. Those who have been very successful should go on to a more challenging ILU or perhaps to an entirely different type of experience. Students who have had limited success may be advised to try selected portions of the ILU again or encouraged to try other activities that might help them achieve the objectives. These alternatives can be assessed and planned through a concluding conference with the student (fig. 4.8). Student opinion should be helpful in evaluating not only the level of progress but also the unit itself. ILUs should be subject to evaluation and revision, and student feedback can be an important element in this process.

Postassessment Card

Now that you have completed this ILU, let's see how you can do on those exercises you tried at the beginning of this unit.

Get the Preassessment Cards and repeat the activities on those cards. Record your answers on the Postassessment Recording Sheet that you will find in the folder on my desk marked "ILU: Volume Measurement." Do these exercises seem easier this time?

Place the recording sheet in your ILU folder and bring the folder to me. I would like to sit down and talk with you about how you did on this ILU and also how you think this unit may be made better.

Fig. 4.7

Fig. 4.8

Preparing the materials package

The key to a successful independent learning unit lies in the quality of the activities represented by the materials package. The materials package should contain all the materials needed to carry out each activity. If an activity card refers students to a puzzle, game, worksheet, or some manipulative material, they should be able to locate it quickly and begin work.

Much of the needed material can probably be found in school. If the ILU is planned and developed far enough in advance, additional commercially available material may be acquired.

Some teachers have found that local businesses, manufacturing firms, and professional offices (e.g., architectural, engineering, and accounting firms) are willing to contribute surplus materials or "imperfect" products that can be used for manipulative activities. Examples of such materials and some ways in which they might be used follow:

Spools from mills—measuring circumference; finding the volume of a cylinder

Small blocks of wood from furniture factories—counting; finding the volume of a rectangular solid

Scrap wood and nails from lumber companies—making geoboards; constructing primitive surveying instruments

Old blueprints—measuring to scale; representing three-dimensional surfaces in two dimensions

The successful acquisition of useful free material is an art that depends on establishing and maintaining good relations with the community. One successful approach that has generated substantial interest and an abundance of materials is to write to business and professional people requesting that they contribute their surplus material for use in the classroom (Abruscato and Hassard 1976). Attaching a "scrounge list" like the one in figure 4.9 will help stimulate the potential contributor's thinking. The list can be tailored to fit the type and range of materials needed.

By including on the list materials other than those potentially usable for ILUs, the teacher may be able to secure more general types of equipment that can assist in the organization of the classroom. Surplus files, cabinets, notebooks, lumber, and cinder blocks (for storage shelves) can all be used by the creative teacher to establish an environment that will foster independent learning.

Managing Independent Learning Units

Preserving information and materials

If activity cards are used, then some attention must be paid to the preservation of the cards. Printed, written, or typed directions can be preserved from accidental erasure or smudging by laminating the cards with a clear plastic coating.

If audiotaped directions are required, it is advisable to use cassettes instead of reels. With the cassette, the playback equipment is small and easily used by children. Moreover, tape cassettes can be prepared so that

SCROUNGE LIST

Storage Materials

Files	Trays	Fruit trays
Cabinets	Gallon glass jars	Packing materials
Small containers	Large plastic containers	Fruit crates
Folders	Shelving	Discarded display racks
Plastic bottles	Cartons	Plastic cups
Dispensing bottles	3-gallon ice cream containers	

Paper Materials

Notebooks	Large sheets of blank	All unusual kinds of
Ledgers	paper	paper (end cuts, damaged,
Light-colored smooth-	Pieces of billboards	or sample)
surfaced cardboard	Large cardboard tubes	File cards

Building Materials and Fixtures

Lumber	Moulding wood	Nuts and bolts
Pipes	Hinges and fittings	Ceramic tiles
Linoleum	Sawdust	Concrete blocks
Tiles	Formica squares	Doweling
	Damaged bricks	

Hardware Materials

Printed circuit boards	Low-voltage light bulbs	Pegboards
Discarded components	Fluorescent light fixtures	Hot plates
Wire	Steel ground rods	Magnets
Excess color wires	Weighing devices	Rubber tubing
	Ball bearings	

Maintenance Materials

Storage batteries	Transfer gears and pulleys	Electric motors (any
Floor brooms and	Battery rechargers	size)
dustpans	Hand air pumps	Wheels

Sample Materials

Sample tile charts	Drapery and upholstery	Wood
Hardware	samples	Linoleum and tile samples
Books	Sample food cans and boxes	Wallpaper books
Sample rug swatches	Color samples	Paper samples

Scrap Materials

Paper scraps	Material scraps	Tile scraps
Wood curls	Leather and lacing scraps	Metal scraps
Scrap plastic (trimmings	Wallpaper scraps	Shavings
and cuttings)	Aluminum scraps	
Rug end pieces		

Miscellaneous Materials

Golf tees	Programmed learning units for
	basic job training

Fig. 4.9

they will not be subject to accidental erasure: simply snap off the two plastic tabs located at the back edge of the cassette *after* the directions have been recorded. This erasure-prevention system can easily be undone by using Scotch tape to cover the spaces that previously contained the tabs.

All the materials for a particular independent learning unit should be stored together. If the unit does not include manipulative materials, then manila folders or clasp envelopes can hold the activity cards and worksheets. A checklist of the materials should be written on the outside of the package. Students should be instructed to check the list against the contents of the folders or envelopes before beginning the unit. Teachers may wish to maintain their own file of extra consumable materials so that an ILU can be quickly replenished.

For units that include manipulative materials, tote trays or tote boxes can be used. Plastic tote trays can be purchased through conventional school-supply companies; however, a much more efficient and inexpensive way to obtain them is to browse through the collections of plastic containers having tight-fitting lids found in supermarkets and department stores. A variety of containers that can be used for ILUs is shown in figure 4.10.

If the school cannot afford plastic trays, shoe boxes can be used. Students and teachers can bring in extra shoe boxes from home, or a local department store might supply leftover boxes.

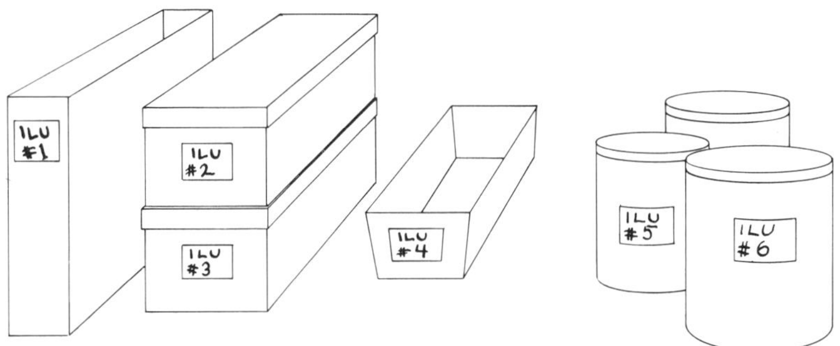

Fig. 4.10. Types of containers that can be used for ILUs

Since shoe boxes are not particularly attractive, students may wish to paint them or cover them with colored plastic adhesive film. A checklist of enclosed materials should be affixed to the lid.

Keeping track of student responses

Because students' data sheets, worksheets, and other materials can easily be misplaced or lost, a file box with a folder for each student should be maintained. The students take out their own folders each day and work on the unit. At the end of the class period, they place their work in the file folder and return the folder to the box. The work can then be periodically reviewed at the teacher's convenience to check student progress.

Maintaining records of student progress

An efficient way for the teacher to keep track of the students' progress is through the use of a master progress sheet for the class (figs. 4.11 and 4.12).

Student Name	ILU #1: Division			ILU #2: Fractions			ILU #3: Decimals			ILU #4: Checkbook		
	Pre	Post	Date of Final Conference	Pre	Post	Date of Final Conference	Pre	Post	Date of Final Conference	Pre	Post	Date of Final Conference
Cindy B.	70	90	9/30	15								
Becky B.	20	80	10/15	35								
John S.	60	90	9/15	70	100	10/1	20	100	10/21	50	90	11/30
Sam D.	60	95	9/26	10								
Judy F.	60	95	10/12	20	80	10/30	30	90	11/25			
Todd H.	70	85	9/20									

Fig. 4.11. Sample of a master progress sheet for a classroom in which ILUs are sequenced for skill development

Student Name	ILU #1: Probability			ILU #2: Surveying			ILU #3: Slide Rule			ILU #4: Brainteasers		
	Pre	Post	Date of Final Conference	Pre	Post	Date of Final Conference	Pre	Post	Date of Final Conference	Pre	Post	Date of Final Conference
Barry B.				30	90	11/1				25	95	10/12
Shirley D.	40	85	10/15									
Jane H.	70	100	2/5				10	90	12/1	70	90	12/20
Steve H.							50	100	11/23			
Warren P.										15	95	12/5
Toby R.				50	80	10/1	20	85	11/1			
John R.										30	80	10/15

Fig. 4.12. Sample of a master progress sheet for a classroom in which ILUs are used for enrichment

The master progress sheet is a valuable source of information for making decisions about selecting appropriate ILUs. By referring to it, the teacher can help students determine which ILUs would be appropriate next. Posttest scores on previous units might indicate whether the ILU is a useful approach for a particular student. High pretest scores may reveal that the prepared units offer insufficient challenge.

Selecting ILUs

As with any other mode of instruction, the ILU approach requires that the teacher carefully select an instructional procedure that is appropriate for a given student. If a sufficient quantity and variety of ILUs are available, then the matching of ILUs to student interests and abilities becomes easier. The following questions may be helpful in facilitating the selection process:

1. Does the student have the prerequisite mathematical knowledge and skills to enter the unit?
2. Does the student have the reading ability to comprehend any written directions?
3. Does the degree of independence required by the structure and sequence of the unit match the student's ability to handle independent work?
4. Does the student have the necessary psychomotor skills to perform the manipulations required in the unit?
5. Will the reinforcement system built into the unit be of sufficient strength to gain and maintain the student's interest?
6. Might other materials available in the classroom be more effective than the ILU in helping the student?

Working with students on an ILU

A major purpose of the ILU is to give students an opportunity to work independently. The natural tendency to "teach" the ILU to the student will interfere with one of its underlying purposes. If the unit is properly designed and within the achievement level of the student, the teacher should be able to function as a *facilitator*—not simply to sit at the desk and correct papers or plan tomorrow's lessons but to encourage and be available to help when help is needed. Teachers should circulate around the room, talking with students who need encouragement and staying out of the way of those who are immersed in their unit and making progress.

Producing sufficient numbers of ILUs

One recurrent frustration of classroom teachers is the lack of sufficient resource materials to establish a rich learning environment. This is a problem regardless of the mode of instruction being used and can be especially critical when teachers try to include independent learning activities. The achievement of students and efficient management will depend on the quality and quantity of independent learning units available. Four suggestions for the production of sufficient quantities of ILUs follow:

1. Each teacher in a school could prepare his or her own ILUs.
2. In-service workshop time could be used to involve other teachers in the production process.
3. Students could assist in the construction of ILUs.
4. Paraprofessionals and preprofessionals could contribute to the production of ILUs.

The particular approach or combination of approaches used will vary from school to school.

Incorporating ILUs in the total instructional program

The uses of any ILU depend on its topic and the reading level reflected in the directions. A unit that provides enrichment experiences at one grade level might be used as a remedial technique for students at a higher grade level (e.g., a unit on place value). This duality of functions can be encouraged through the sharing of ILUs across as well as within grade levels.

ILUs can clearly serve a special enrichment purpose for advanced students who are working beyond the curriculum or text material. Such ILUs can extend the students' thinking and skills into areas not generally considered at that grade level. For example, a unit on topology could be prepared for upper elementary or junior high school students.

The ILU format could serve as an alternative way of presenting a topic that is an integral part of the curriculum. This use provides the students with a good change of pace in the method of instruction. Examples of topics that can be presented as ILU alternatives are telling time in second grade, adding fractions in fifth grade, and graphing linear equations in first-year algebra.

The ILU can also serve as an integrator of basic skills. Skills that are generally taught in a linear way can be tied together through a well-crafted unit that focuses on a specific project, such as an ILU on check writing that ties together addition, subtraction, and multiplication skills.

Some Closing Thoughts

Although the ILU approach differs to some extent from more conventional modes of instruction, it is based on a synthesis of some commonly held beliefs about the nature of learning and the instructional process. The uniqueness of the ILU approach lies in the manner through which it integrates such pedagogical techniques as instructional objectives, preassessment and postassessment devices, provision for a knowledge of results, and teacher encouragement.

If students can find success within the ILU approach, then they will be ready to go on to forms of independent learning that provide continually less direction and dominance from the teacher. The teacher becomes more and more of a guide who is available to assist when help is needed without overpowering the student's own creative ideas and energies.

Developing an independent learning unit can be a creative professional experience for teachers and can provide an opportunity to explore and respond to a variety of questions relating to mathematics education: "What needs to be learned?" "What are appropriate learning activities?" "How can I provide for the wide range of achievement levels in a class?" Developing and using ILUs can help each of us respond to these questions.

REFERENCES

Abruscato, Joe, and Jack Hassard. "Acquiring and Using Resources." In *Loving and Beyond*, pp. 34–45. Pacific Palisades, Calif.: Goodyear Publishing Co., 1976.

Cunningham, James B. *Teaching Metrics Simplified*. Englewood Cliffs, N.J.: Prentice-Hall, 1976.

Higgins, Jon L., ed. *A Metric Handbook for Teachers*. Reston, Va.: National Council of Teachers of Mathematics, 1974.

Johnson, Rita B., and Stuart R. Johnson. *Assessing Learning with Self-Instructional Packages*. Reading, Mass.: Addison-Wesley Publishing Co., 1973.

Mager, Robert F. *Preparing Instructional Objectives*. Palo Alto, Calif.: Fearon Publishers, 1962.

Popham, W. James, and Eva L. Baker (a). *Planning an Instructional Sequence*. Englewood Cliffs, N.J.: Prentice-Hall, 1970.

———— (b). *Systematic Instruction*. Englewood Cliffs, N.J.: Prentice-Hall, 1970.

Schoen, Harold (a). "Self-Paced Mathematics Instruction: How Effective Has It Been?" *Arithmetic Teacher* 23 (February 1976): 90–96.

———— (b). "Self-Paced Mathematics Instruction: How Effective Has It Been in Secondary and Postsecondary Schools?" *Mathematics Teacher* 69 (May 1976): 352–57.

Wilderman, Ann M., and Harold S. Resnick, and David E. Kapel. *Metric Measure Simplified*. Boston, Mass.: Prindle, Weber & Schmidt, 1975.

5

Organizing Goal-referenced Instructional Units

Robert F. Nicely, Jr.

*T*he alternative described in this essay is built on several basic assumptions that come from the concepts of diagnostic-prescriptive teaching and mastery learning. Diagnostic-prescriptive teaching assumes that a set of expectations (objectives) for students can be clearly delineated, that appropriate instruments or procedures can be developed to assess achievement of those expectations, and that resources are available to help those students who do not have the competence specified in the expectations. The concept of mastery learning is based on the work of Bloom (1968), who asserted that "most students can master what we have to teach them, and it is the task of instruction to find the means which will enable our students to master the subject under consideration. Our basic task is to determine what we mean by mastery of the subject and to search for the methods and materials which will enable the largest proportion of our students to attain such mastery."

The Instructional Model

The model that follows is a plan for developing and implementing instructional units. Each unit is based on a set of behavioral objectives, some of which students must master and others of which are optional at the discretion of the teacher or curriculum committee of the local school district.

Each unit includes behavioral objectives, diagnostic procedures designed to assess each student's achievement of the required objectives, and a variety of instructional strategies and materials related to the behavioral objectives. Record-keeping forms and a plan for the organization and management of the classroom that includes flexible grouping within the class are necessary for the successful implementation of each unit. (Illustrations of these components and plans are placed appropriately throughout the essay.)

The model does not require continuous progress across grade levels or units but rather allows students to work on the optional objectives and activities once they have demonstrated a mastery of the locally determined required curriculum. Experience has shown that the model does not require additional personnel but is quite manageable for one teacher having about thirty students (Nicely and Shepler 1971). It is designed to increase the degree of individualization that can occur within such a classroom setting by grouping and regrouping students on the basis of the results of locally developed diagnostic procedures. The model can be adapted to the resources that are available in any given classroom, whether that classroom contains only minimal instructional resources or whether it contains a relatively rich materials environment, including multimedia materials and more than one textbook. Provision is made for most students to attain a mastery of the locally determined minimum required curriculum, and teachers are enabled to provide systematic reinforcement of this basic curriculum. A variety of ways for students to apply the basic content is provided, and students are involved in making some of the decisions. Alternative objectives are available to those students who have mastered the required curriculum.

Curriculum Development

Since the model is based on the development and implementation of instructional units, the first task is to identify and develop those units. Figure 5.1 shows the steps many elementary and secondary school teachers have used in developing a unit.

Identifying units and objectives

Since textbooks provide the basis of the mathematics curriculum for most students in this country, many teachers find it convenient to organize instructional units around textbook chapters. At the secondary level, the textbook chapters are usually built around a single theme. Generally, a concept is treated intensely in a chapter, and the chapters are arranged in a sequence from simple to complex. Often the objectives within a chapter can be arranged in a linear sequence.

At the elementary level, a similar organization is often found, but in addi-

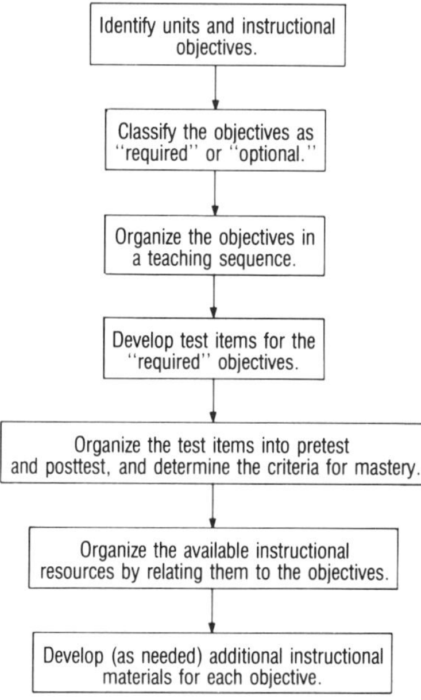

Fig. 5.1. Steps in developing a unit

tion, it is not unusual for a textbook chapter to include several themes not closely related. Usually, the authors of such texts have strived to spiral the mathematics content from simple to complex throughout the book. A chapter might contain a variety of objectives built on less complex objectives from the same concept areas found in a preceding chapter. The objectives in one chapter are usually prerequisites to more complex objectives in the same concept area in succeeding chapters. The plan for classroom management described in the next major section of this essay can be used regardless of the way objectives within a unit are organized.

Once the theme of the unit has been identified, the teacher's next task is to determine the instructional objectives. In most contemporary textbooks, objectives are included in the teacher's edition. (The instructional unit illustrated in this essay was developed from a chapter in *Exploring Elementary Mathematics* [Keedy et al. 1970].) If the objectives provide adequate guidance, as perceived by the teachers, that is all that is needed. However, if the objectives are missing or are inadequate according to the teacher's professional judgment, there are excellent resources that can be used for aid in identifying, writing, or modifying them. Mager (1962) and Kibler, Barker, and Miles (1970) offer procedures and criteria for writing and clarifying objectives. Nicely (1970) specifies procedures for identifying the objectives in textbooks where the authors have not written them.

Classifying objectives

All mathematics teachers at one time or another are faced with the "coverage" quandary. Many feel an obligation to "cover" the textbook. What they fail to realize is that a textbook is written to provide more than enough information for the best students and that it is unreasonable to expect all children to attain a mastery of *all* the information contained in a given textbook in the same amount of time. Therefore, it is critical for teachers to set priorities for the teaching of the identified objectives.

Once the objectives have been identified, the teacher should examine all of them for each instructional unit and then select those that he or she feels *all* students should master. These objectives would be classified "required." The remaining objectives would be "optional." The optional objectives are often those that "I'd like to teach if I had time."

This approach to determining the required and optional curriculum usually helps to alleviate the problem that normally occurs every school year in May when the teacher in a self-contained classroom realizes that only a few weeks remain and there are too many pages left in the book. Obviously, conversations with teachers from the contiguous grade levels concerning curriculum articulation would be helpful.

Organizing objectives

Once the objectives are classified as required or optional, the next task is to organize them for instruction. If all or some of the objectives in the unit are closely related, it is often useful to arrange them in a linear sequence. In such a sequence, the mastery of one objective is necessary before the student can succeed with the next objective. Figure 5.2 shows such a

---

1. The student demonstrates a mastery of the basic subtraction facts in a rapid-response drill, both orally and written.

2. The student demonstrates a mastery of subtraction computation involving numbers with at least two digits and not requiring regrouping.

3. The student demonstrates a mastery of subtraction computation involving two-digit numbers and requiring regrouping at the tens place.

4. The student demonstrates a mastery of subtraction computation involving three-digit numbers and requiring regrouping at the tens *or* hundreds place.

5. The student demonstrates a mastery of subtraction computation involving three-digit numbers and requiring regrouping at the tens *and* hundreds places.

Fig. 5.2. Illustration of objectives in a linear sequence (adapted from the D-Subtraction unit in *Individually Prescribed Instruction—Unit Manual* [Miller and Nicely 1970])

sequence. If some of the objectives in the unit are not closely related to the others, then the teacher can have more freedom in deciding where to place them in the instructional sequence.

Figure 5.3 illustrates a page from the table of contents for the sample unit from *Exploring Elementary Mathematics, Book 3.* The teachers who organized this unit developed a coding system to keep track of the sequence of objectives, the testing program, and the instructional resources. Each objective in the mathematics curriculum has a unique five-digit code: the first digit identifies the grade level, the next two digits identify the unit, and the last two digits identify the order of the objectives in the teaching sequence. For example, the code number 30602 indicates that the objective is taught in third grade as part of the sixth unit, and it is the second objective in that unit. In unit #306, the teachers felt that the objectives listed by the textbook authors were adequate for their needs and therefore decided

Grade level/Time   3/Midyear   **Unit Name:**   Multiplication and Division

Unit Number: 306

Section	Page
I. Table of Contents and Objectives	1
II. Pretest	2–3
III. Instructional Resource Guides (required objectives)	R-1 through R-5
IV. Instructional Resource Guides (optional objectives)	Op-1
V. Posttest	last 2 pages

Objective Code No.	Behavioral Objectives	Required	Optional
30601	Demonstrate a mastery of the basic multiplication facts (with 1, 2, 3, and 5 as factors) and the related division facts in rapid-response drill or testing, both written and oral.	X	
30602	Demonstrate an understanding of the relationship of multiplication to division by writing related sentences.	X	
30603	Use the commutative property and the distributive property of multiplication over addition to discover new facts.	X	
30604	Distinguish between the terms "factor" and "product."	X	
30605	State generalizations about a product when 1, 0, or 10 is a factor.	X	
30606	Solve word problems involving multiplication and division problems and label answers.		X

Fig. 5.3. Table of Contents page for sample unit

not to modify the wording or the ordering of the objectives. However, they did decide to add an optional objective (#30606) to the unit.

Developing test items and tests

After the objectives have been arranged in an instructional sequence, the next task is to develop test items for the required objectives. If the test items are developed for the required objectives only, the time spent in testing is held to a minimum. Such a testing program provides information about the competence of students only in reference to the minimum curriculum. There is no real need to test the students for a mastery of the optional objectives. However, teachers should develop alternative ways to recognize the students' accomplishment of the optional objectives as well as ways to provide feedback about their performance.

Figure 5.4 illustrates part of a sample pretest for the illustrative unit of five objectives. The first set of items, 30601, tests the student's knowledge of the first objective, that is, basic multiplication and division facts. The rectangular record chart in the upper right-hand corner of the test shows

Unit 306   Pre/Posttest

Name ____

Objective	Tot.	Min.	Rx
30601	20	17	
30602	6	5	
30603	5	5	
30604	3	3	
30605	3	3	

30601

$$\begin{array}{cccc} 3 & 2 & 5 & 2 \\ \times 4 & \times 2 & \times 8 & \times 6 \end{array} \quad \text{etc.}$$

$$24 \div 8 = \square$$
$$15 \div 3 = \square \qquad \text{etc.}$$

30602

Write two related division sentences . . .
(a) $3 \times 5 = 15$ ___________   ___________
(b)   etc.

30604

$5 \times 8 = 40$
Draw a circle around the factors.
Draw a box around the product.

Fig. 5.4. Partial sample of a pretest

that the teachers who developed this test had a total (Tot.) of twenty items for objective 30601 and felt that seventeen was the minimum number (Min.) a student must get correct in order to demonstrate mastery. Objective 30604 requires the student to identify the factors and the product when given a multiplication sentence. Here there are three required responses, and the teachers felt that the student must get all of them correct in order to meet the minimum acceptable level of competence. The record chart also contains a column labeled "Rx." If a student mastered an objective on the pretest, the teacher wrote P in the appropriate space; if the student did not master it, the teacher left the space blank.

In the process of developing test items, a balance must be attained between having an adequate number of test items (to sample the behavior so that a relatively accurate decision can be made about each student's competence) and keeping the test short enough to allow the students to complete it in one class period or less. The decision on the minimum acceptable level of mastery is somewhat arbitrary but tends to range from 75 to 100 percent, depending on the type of objective and the number of items. If a student can do three out of four problems of a given type correctly, it is generally safe to assume minimal competence on that objective. However, teachers should still be interested in trying to determine why the other question was answered incorrectly. A useful resource for teachers to use when writing test items for pretests and posttests is a book entitled *Preparing Criterion-referenced Tests for Classroom Instruction,* by Norman Gronlund (1973).

Organizing instructional resources

Once the objectives have been identified and classified and the tests developed, the next task is to organize the instructional resources by relating them to the specified objectives. Figure 5.5 illustrates a sample instructional resources guide. Usually, one such page for each objective in a unit is adequate. The same form can be used to organize resources for either required or optional objectives. The left side of the guide contains a statement of the objective and an example of the test item for it. The instructional resources are classified in three categories: Basic, Applied Learning, and Enrichment.

The *basic* instructional resources require the teacher to play a major role in presenting and directing the learning activities. Generally, these resources are the ones that would be used if the teacher was largely going to ignore individual differences and teach the whole class under the assumption that all the students were at the same stage of readiness and mastery. Typically, these resources are comprised of textbooks, teacher-prepared worksheets, and visual aids to assist the teacher in clarifying lectures or discussions.

Unit <u>306</u>   Required Objective #<u>30602</u>

or

Optional Objective # _______

| Objective | Instructional resources | | |
	Basic	Applied Learning	Enrichment
Statement Demonstrate an understanding of the relationship of multiplication to division by writing related sentences. *Example* Write two related division sentences for each multiplication sentence. $8 \times 1 = 8$ ___ ___ $5 \times 7 = 35$ ___ ___ $3 \times 4 = 12$ ___ ___	Holt, Rinehart, and Winston— Bk. 3, 1970 pp. 193, 194 #1–8 p. 195 #1–15 number line counting disks flash cards	Holt, Rinehart, and Winston— Workbook 3 p. 69 #1–15 Worksheet 30602 #1* Worksheet 30602 #2* Worksheet 30602 #3 *Audiotape 30602 #1 is coordinated with these worksheets. Domino game #30602	Project 30602 #1 Project 30602 #2

Fig. 5.5. Sample instructional resources guide

The instructional resources for *applied learning* usually require the teacher to initiate the activity, allow the students to progress independently at their own speed for a relatively short period of time (fifteen to thirty minutes), and then bring closure to the activity. These resources are either for those students who have almost mastered an objective but need some reinforcement or for those students who have demonstrated readiness for learning an objective and have a high probability of being able to manage their learning with little personal direction from the teacher. As implied in figure 5.5, two useful types of resources for applied learning are (1) audiotapes that are related to worksheets or textbook pages, and (2) games. Such activities can be used by individuals or small groups at learning stations. A sample of one type of such a resource for objective 30602 in the illustrative unit is given in figure 5.6.

The instructional resources for *enrichment* require relatively little personal direction from the teacher. They are often designed to enable students who have mastered the objective to (1) work in nonrequired curriculum areas, both within and outside mathematics; (2) conduct research in the school library; (3) assist other students, at the same or different grade levels; and (4) develop games, learning stations, or other materials for use by other students. The enrichment activities and resources can be designed to allow students considerable freedom in using resources as they deem appropriate. A sample of one type of enrichment resource that was developed to accompany objective 30602 is given in figure 5.7.

Developing additional materials

The final and ongoing step in this process of developing a unit is for teachers to produce, as needed, additional instructional materials for each objective. Whenever any of the three categories of instructional resources in figure 5.5 is blank or has relatively few resources, the teacher should give high priority to developing or acquiring additional materials. Such an organizational process allows teachers to examine materials selectively with an eye toward choosing those that will help them meet diagnosed curriculum weaknesses.

Typically, the supply of instructional resources in the basic category is more than adequate. A textbook and related aids are common in most instructional settings. However, within the categories of applied learning and enrichment, the quantities of resources and related activities are often insufficient. It is here that teachers must usually focus their efforts to develop instructional products. Figure 5.6 shows an illustration of a teacher-developed resource for use by an applied-learning group in conjunction with objective #30602.

Domino Game #30602

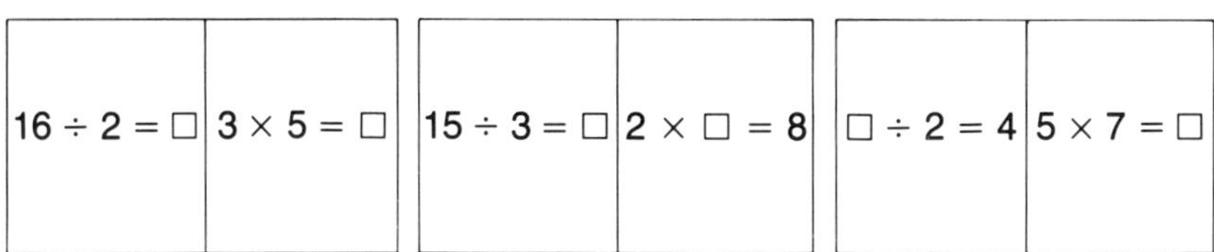

NOTE: Prior to having students work with "Domino Game #30602," the teacher had conducted a whole-class session early in the school year to teach them how to play dominoes. The domino format has proved to be quite useful because once the students learn the rules and procedures, they can play the game with little or no direct supervision from the teacher. Dominoes can be played by a small group or by an individual student. The pieces for the games can be kept in shoe boxes or some similar container. The container is *labeled with the appropriate objective code number.*

Additionally, the teacher or students can generate many forms of the game involving specific objectives using 7.5 cm × 12.5 cm cards and magic markers. The task of generating games is one of the possible activities for students assigned to enrichment. Another enrichment activity would be for that student to initiate and supervise the playing of the domino game by the applied-learning students.

Fig. 5.6. Sample instructional resource for applied learning

As mentioned earlier, another type of resource for applied learning is the audiotape/worksheet activity. Many teachers find that students prefer audiotapes that are made by their own teacher. Such tapes have a familiar voice and are usually consistent with the teacher's own style and timing for giving information and directions. One method that seems to work well for teachers who make audiotape/worksheet activities is to (1) design the worksheet first, (2) rehearse, in private, what would be said to the students when working with them in a face-to-face situation using the worksheet, and then (3) record, in private, the rehearsed verbal directions and comments, allowing adequate time for the students to respond. For those situations where an extended amount of time is needed for students to complete an assigned task, the teacher might choose to tell the students to turn off the tape until they have completed the task. Then they can turn the tape back on to get answers, further directions, and so on. The Wollensak Mathematics Teaching Tapes program (1968) is a model that is appro-

---

Enrichment Project 30602 #1

1. Get a blank piece of paper and draw a chart on it that looks like this. Make yours bigger.

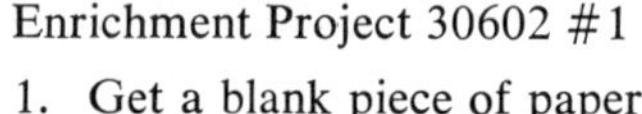

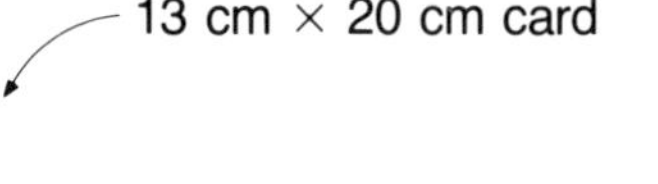

×		5		4	← factors
				20	⎫
2	0		2		⎬ ← products
				16	⎭
3					

factors

2. This is a **multiplication** chart. *Some* of the factors are given, and *some* of the products are given. You have to find the missing factors and products. Fill in the chart you drew.
3. After you have finished, check your chart with the teacher's chart on the back of this card.

NOTE: By placing the objective code (and any additional identifying codes) on the nontextbook instructional materials, teachers can classify, store, and retrieve the materials with ease. In fact, students can be taught to retrieve (and return) selected materials. Without such a coding system, materials can become lost or underused.

Fig. 5.7. Sample enrichment instructional resource

priate for teachers to examine as they make their own audiotape/worksheet activities.

Figure 5.7 illustrates a teacher-developed resource for use by an enrichment group in conjunction with objective #30602. Several additional possibilities for enrichment students were mentioned in the notes in figure 5.6. As mentioned earlier, enrichment tasks and resources do not need to be limited to the mathematics curriculum. In addition, teachers should not assume that enrichment activities should *only* be inquiry oriented or involve tasks designed to foster student involvement at the complex end of the cognitive taxonomy (Bloom 1956). Enrichment tasks can be oriented to acquiring knowledge and skill, just as inquiry strategies can be used at the lower end of the taxonomy and in conjunction with teacher-directed and teacher-guided instruction associated with the basic and applied-learning resources.

Classroom Organization and Management

In this process of developing and implementing instructional units, the next steps are to examine classroom organization and management.

Physical setting and record keeping

Figure 5.8 illustrates a classroom arrangement developed by one teacher

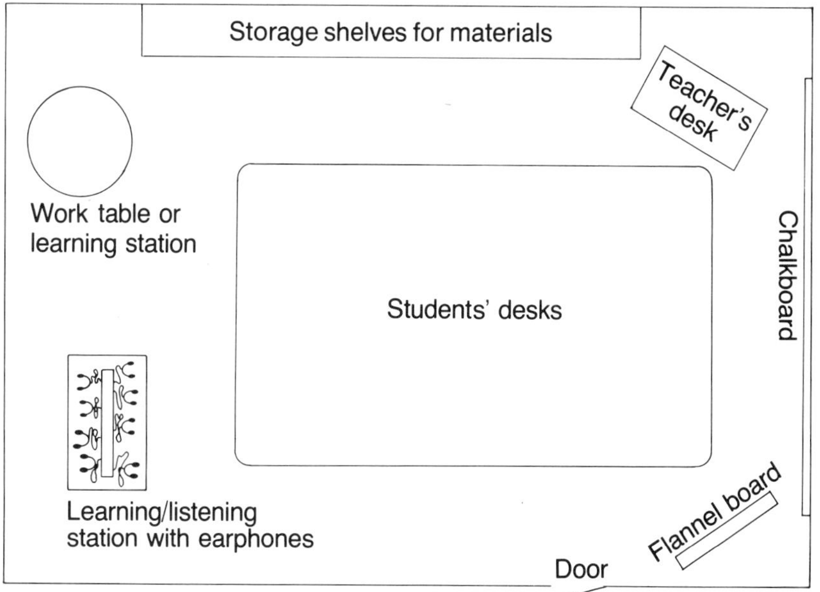

Fig. 5.8. Sample room arrangement

to accommodate three instructional groups. Obviously, each teacher must design a plan for organizing the classroom that takes into account the physical structure of the room and the furniture within it.

In the room illustrated, students at their desks often used the basic instructional resources and received much teacher direction. Students using applied-learning and enrichment resources worked on activities at the

Unit Code **306**   Student Record Form

Key:
- **P:** Mastered on pretest
- **/:** Working on objective
- **X:** Mastered after working on it

		Objectives																	
		Required										Optional							
		30601	30602	30603	30604	30605						30606	30607						
1	Student A	P	P	P															
2	Student B	P		P															
3	Student C	P		P															
4	Student D	P	P	P	P	P													
5	Student E	P																	
6	Student F	P	P	P															
7	Student G																		
8	Student H	P	P	P	P														
9	Student I	P	P																
10	Student J																		
11	Student K																		
12	Student L																		
13	Student M	P																	
14	Student N																		
15	Student O	P			P														
16	Student P																		
17	Student Q	P																	
18	Student R		P																
19	Student S																		
20	Student T	P																	
21	Student U	P																	
22	Student V	P	P																
23	Student W																		
24	Student X																		
25	Student Y			P															

Fig. 5.9. Sample student record form

listening/learning station, the learning centers, the work tables, or at their own desks, depending on the task and the available resources.

Once general decisions about the physical arrangement of the room have been made, attention must be focused on grouping and managing the students. Figure 5.9 illustrates a student record form for a class, with the results of a pretest entered. The record form provides for the objectives to be listed as required or optional. Such a record form (one for each unit and class) allows the teacher to tell at a glance which students have mastered which objectives on the pretest and on which objectives each student needs to work in order to attain mastery.

Flexible grouping

The flowchart in figure 5.10 illustrates a plan that allows the teacher to move the entire class through a unit from objective to objective. Although it may appear that the teacher is teaching about one objective in isolation from others, it is not so—usually, each objective is related to other objectives, and the teacher draws these relationships while *focusing* on one objective at any given time.

The flowchart illustrates the diagnostic-prescriptive processes of (1) pretesting; (2) using the results of the pretests to place students in one of three instructional groups; (3) providing instruction for those groups; and then (4) bringing closure to the lesson, making provisions for rediagnosis and the recycling of students as necessary. Teachers using this plan of instructional management need only the student record form (fig. 5.9) to keep track of which students have mastered which objectives.

For example, on the first day of instruction (using the information in fig. 5.9), students G, J, K, L, N, P, R, S, W, X, and Y would be assigned to the basic instructional group. The remaining students would be assigned to either the applied-learning group or the enrichment group, depending on the teacher's judgment of each student's competence and the availability of instructional resources. At the end of the lesson, the teacher would know whether most students in the basic group had demonstrated adequate knowledge of objective 30601 and would then decide whether to continue that lesson the next day or move to a new lesson. This process would continue throughout the unit as indicated in figure 5.10. The resources for the three potential groups are listed on the instructional resources guide in figure 5.5. This plan enables students to work on optional objectives and activities throughout a unit if they have demonstrated a mastery of the required objective that is to be treated on any given day. Many teachers like this plan because it allows them to focus on one topic at a time and still provide alternative activities for those students having different needs relative to that topic.

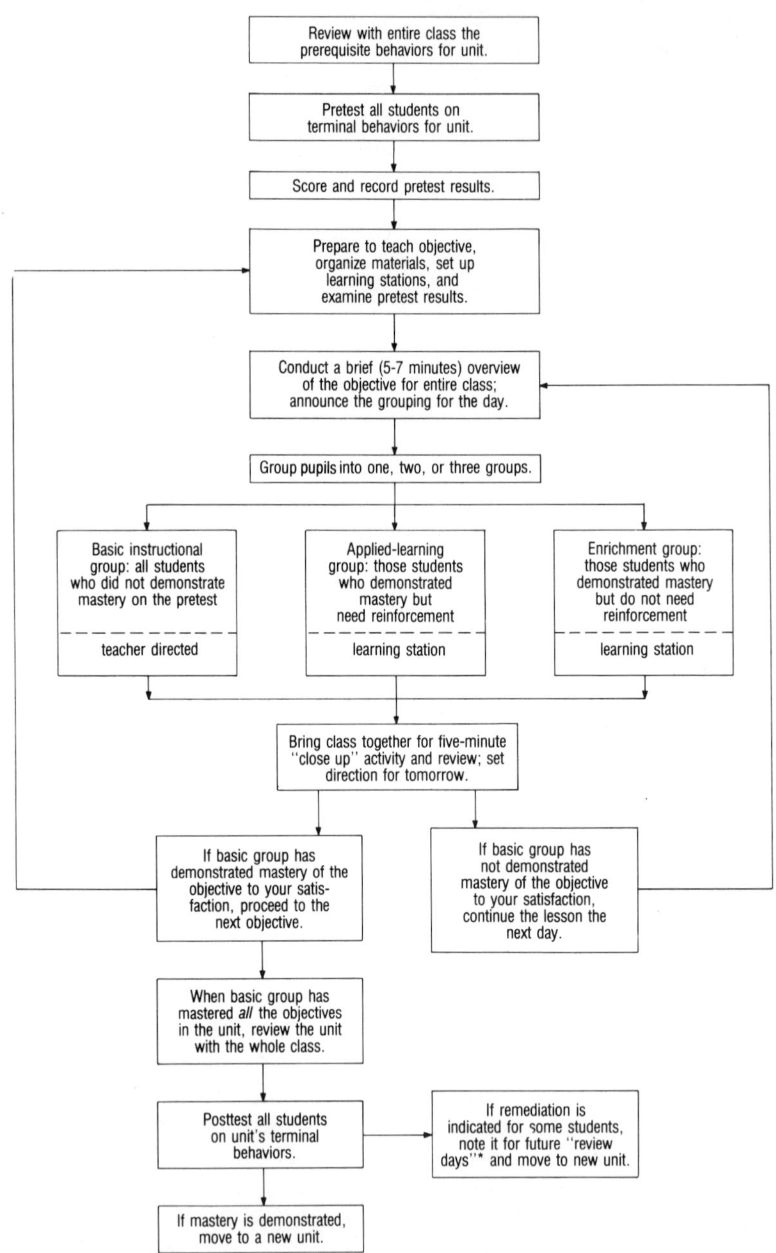

*Sometimes it is desirable to have a whole-class review of the major concepts and skills that have been developed up to that point. This "review day" might be conducted once a month and is a good place to offer remedial help, use peer tutors, and so on.

Fig. 5.10. Instructional flowchart for flexible grouping

Although this plan is appropriate for use both in units whose objectives can be arranged in a linear sequence and in units whose objectives are not closely related, a refinement of the plan can be made for those units in which the objectives can be arranged in a linear sequence. This modified plan (see fig. 5.11) permits students to progress throughout the unit at their own rate of learning and then move on to optional objectives as time and resources permit.

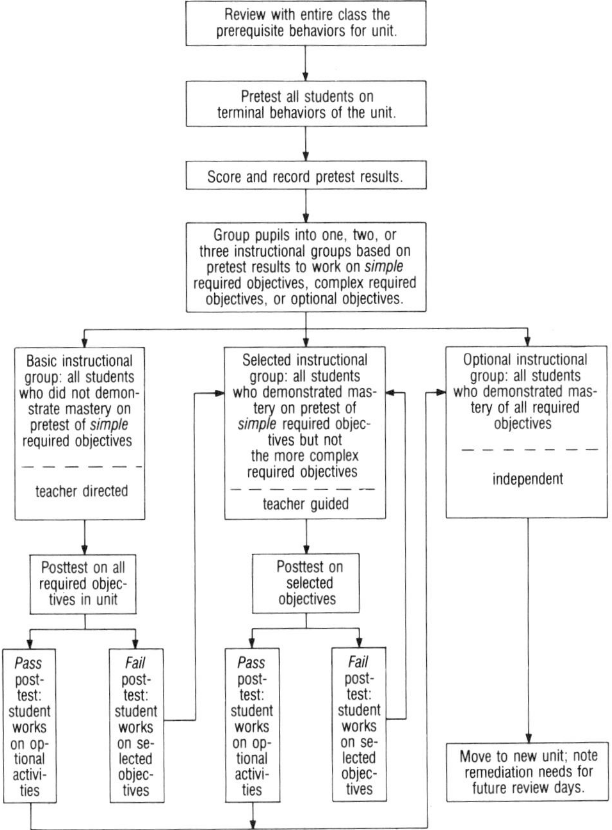

Fig. 5.11. Optional instructional flowchart for units having a linear sequence of objectives

In order to make this modified system of classroom management operational, teachers must separate the required objectives into two sets: those that are more complex, and those that are relatively simple or that are prerequisite for the more complex ones. This modified plan assumes (1) that a student who does not demonstrate a mastery of the relatively simple required objectives needs personal help from the teacher in acquiring that competence, and (2) that a student who demonstrates a mastery of the relatively simple required objectives is ready to study the more complex objectives in the unit and may be able to study and master them with relatively little direction from the teacher.

The students who do not master the simple objectives (in other words, show zero or near zero competence in the unit) are placed in the basic instructional group. Those students who demonstrate partial mastery of the unit (by mastering the simple objectives) are placed in the selected instructional group. Those few students who demonstrate mastery of *all* the objectives are placed in the optional instructional group.

The same instructional resources guide (fig. 5.5) can be used in working with these groups. The basic resources would be used for the basic instructional group, and the applied-learning resources would be used by the students who were placed in the selected instructional group. The students in the optional group would use the enrichment learning activities.

In order for this plan to work, it is critical that the "Applied Learning" and "Enrichment" columns have adequate instructional activities. As the students in the basic instructional group gain competence and demonstrate their mastery of the basic required objectives, they will move into the more complex required objectives. At that time, if it is instructionally efficient, they can be taught the more complex required objectives in a group setting.

Teachers choosing to use this refined system of management would probably need an additional chart to note the daily regroupings of students based on the progress they made in the instructional period. The student record form is useful for long-range planning, and if it is used with the plan outlined in figure 5.10, it is sufficient for planning daily regrouping.

Summary

The organization and the implementation of goal-referenced instructional units result in a greater degree of individualization than that which occurs in classrooms where whole-class teaching is the usual mode of instruction. The diagnostic-prescriptive process has the practical effect of reducing the class size for the teacher by identifying those students who have relatively little need for direct supervision and allowing them to work at self-instructional or independent tasks. The teacher is then able to give more personal instruction to the remaining students, who have demonstrated that they have a need for such attention.

REFERENCES

Bloom, Benjamin S. "Learning for Mastery." *Evaluation Comment* (May 1968): 1–12.

Bloom, Benjamin S., ed. *Taxonomy of Education Objectives, the Classification of Educational Goals: Handbook 1, Cognitive Domain.* New York: David McKay Co., 1956.

Gronlund, Norman E. *Preparing Criterion-referenced Tests for Classroom Instruction.* New York: Macmillan Co., 1973.

Keedy, Mervin L., Leslie A. Dwight, Charles W. Nelson, John Schluep, and Paul A. Anderson. *Exploring Elementary Mathematics 3.* New York: Holt, Rinehart & Winston, 1970.

Kibler, Robert J., Larry L. Barker, and David T. Miles. *Behavioral Objectives and Instruction.* Boston: Allyn & Bacon, 1970.

Mager, Robert F. *Preparing Instructional Objectives.* Belmont, Calif.: Fearon Publishers, 1962.

Miller, George R., and Robert F. Nicely, Jr. "Individually Prescribed Instruction—Unit Manual." Mimeographed. Pittsburgh, Pa.: Learning Research and Development Center, University of Pittsburgh, 1970.

Nicely, Robert F., Jr. "Development of Procedures for Analyzing Materials for Instruction in Complex Numbers at the Secondary Level." (University of Pittsburgh, 1970.) *Dissertation Abstracts International* 31A (November 1970): 2262.

Nicely, Robert F., Jr., and Warren D. Shepler. "PRIMES Instructional Model for Flexible Grouping—Final Report." Mimeographed. Harrisburg, Pa.: Pennsylvania Department of Education, 1971.

Wollensak Mathematics Teaching Tapes. St. Paul, Minn. 3-M Co. Mincom Division, 1968–76.

6

Organizing a Learning Cooperative: Survival Groups

Herbert Fremont

*E*xamples are numerous in our society of people getting together to accomplish what none could have done alone. In order to live better, many people buy into apartment co-ops. Farmers get together in cooperatives to gain a fairer share of profits. Cooperative food stores are plentiful, and consumer buying services, also a co-op, are on the increase. In each example the members of the co-op derive benefits not otherwise attainable. Why not follow this model in school?

Starting a Learning Cooperative

A learning cooperative is based on the concept of sharing much more responsibility with the student. The familiar pattern of "teacher up front, students out there" must be altered. We learn best by doing, whether it is learning how to drive a car, learning mathematics, or learning how to take responsibility. The learner, not the teacher, should be at the center of the experience.

If teachers are not to take their accustomed role, what will they do? A description of how one class was organized as a learning co-op might clarify this approach.

We enter the learning co-op classroom and see several small groups of students working in the room with one group at the chalkboard. The teacher is walking around, stopping from time to time to talk to a student or to a group of students. Several students are taking tests in the midst of this activity. We see that a rich variety of materials, from books to manipulative devices, is being used or is available on nearby shelves. Several questions come to mind:

- How is the class arranged?
- How do the students know what to do?
- How does the teacher keep track of everyone?
- How are tests administered?

The basis for the scene we observed is what are called "survival groups." If we think back to the days when we were students of mathematics, we can probably recall times when we met regularly with our classmates outside of class to work together on assignments or to try to reconstruct work done in class. Such a study group has helped many college students "survive" in mathematics classes. This experience is not an uncommon one, and an interesting thing has been observed in the study groups: a group of four students can often achieve together an understanding of concepts that none could have accomplished alone. The sum total of the group's work is frequently greater than the sum of what could have been produced by members on their own. The interactions that take place between group members seem to add an entirely new dimension to what the individual members could accomplish. It should not be surprising that much of the mathematics we learn is learned outside the mathematics classroom. Instead of leaving such an important determinant of learning to chance, why not build "survival groups" into our classroom structures? This idea is at the heart of the learning cooperative. How shall we start?

The beginning is made before entering the classroom. We collect as many textbooks, workbooks, aids, games, and so on, as we can that deal with the content of the course. Then we organize this material along content lines to be quite clear about what is available for student use. It is important whenever possible to acquire materials that either have accompanying answers or are self-checking. This will save valuable time for the teacher.

After materials have been collected, we turn our attention to organizing the course. One convenient approach is to use the chapter headings of the class textbook as a vehicle for dividing the curriculum into units. In this way we have a breakdown of the subject that is consistent with student materials.

It is generally best to make changes in customary classroom procedures gradually. Therefore, we begin the class as usual, with some minor but important changes. We distribute textbooks, make presentations to the class as a whole, assign homework, and set the pace for learning for all students by fixing the time period of the first unit and the date for testing on it. Some slight modifications are introduced at this point: we *consult with the students* about what will happen in class. This is done, not to allow them to make decisions (that will come later), but to offer them a chance to express an opinion. For example, we may ask them to consider these questions:

1. Would homework be helpful today?
2. Should more time be spent on this concept?
3. Are we ready for a test in the next few days?

This helps create from the start the kind of learning atmosphere that will lead to students having a greater share of responsibility later.

When the first unit has been completed and tests taken, a natural breaking point with past procedures will be at hand. The fact that test scores are usually well distributed poses a dilemma for the teacher and the students:

- Some students did poorly. Shall we go on to new work?
- Some students did very well. Shall we ask them to wait?
- What about those in between?

Survival Groups

A class discussion on the questions above inevitably leads to a suggestion that somehow we take care of all students by allowing those who are ready to go on to do so and by allowing those who need more time to have it. But how shall this be done? At this point the survival groups in particular, and the learning cooperative in general, may be introduced. The idea of working in small groups may come from the students or it may be suggested by the teacher, but it should be put forth, explained, and discussed. It should be emphasized that such a group is a voluntary coming together of people to help each other learn mathematics. The criteria for forming such groups should be outlined and made clear to all students:

- Some students decide to work with friends.
- Some decide to seek out students with common goals.
- Others seek out students with common interests.
- Still others seek out the better mathematics students.

The arrangement of students into survival groups is a critical point in the development of the learning co-op. If the teacher's attitude at this point leaves the student with little real choice about which group to join—either explicitly through direction or implicitly through nonverbal cues—there will be difficulties throughout the year. If we are to build a situation within which students learn to take the responsibility for their own learning, then we must offer them a genuine choice. If this apparent opportunity is in reality a chance to choose only what the teacher had in mind all along, the learning co-op is being built on a false foundation and will eventually collapse. Choices when offered must be honored. If some students choose to work alone rather than become part of a group, this choice should be respected. Later, we may well try to encourage (not coerce) these students to join one of the survival groups so that they and their classmates may derive mutual benefits, but the final choice must remain with the student.

Before continuing, a digression is in order. The discussion here assumes that a full-time learning cooperative is the objective. It is possible, however, for teachers to begin on a part-time basis. This may be done in two ways. One way is to set aside a day or two each week as "co-op" days, on which the procedures and activities described here would be undertaken. Students could then move into the co-op situation for part of the week and enjoy traditional instruction the remainder of the week. Eventually, one method or the other will probably dominate; however, as a trial for the cautious teacher and as an introduction for the students, one or two days a week may be a satisfactory beginning for the learning co-op. A second way is to offer traditional instruction for the introduction of new topics, letting the class become a co-op when students begin to explore ideas for themselves or to practice the material. This approach also provides a gradual transition from traditional instruction to the survival groups and a full-fledged co-op.

Before being sent off to work in survival groups, the students must clearly understand their responsibilities and the procedures to be followed. In a class of thirty-five, there may be as many as eight groups ranging in size from three to six students, and some students may desire to work alone. If all questions about procedures are discussed as they arise, each student will have a clear picture of her or his role in the operation of the class.

Student Choices

The syllabus for the course is accepted as the minimum mathematics that students are to master, but it is in no way a mandate on how a teacher shall teach, the order in which topics shall be taught, or the length of time to be spent on a given topic. These are all teacher options. If the survival groups are to have responsibility, it will be in those areas where the teacher holds the responsibility. The students are thus limited to selecting topics for study from those contained in the syllabus before choosing any other topics. Within this range, however, they are free to select any unit they desire. Even though mathematics is a highly sequential subject, no one can say what the best possible sequence is for every student. For instance, notice how the order of topics in a course varies from text to text. If textbook authors can order topics in accord with personal preference, why can't our students? Remember, we are trying to capitalize on their self-motivation and desires, both of which may be considerably enhanced by allowing them to select topics of interest to them. In this way, students can arrange a sequence that is suitable for them, even though it may be that no two groups will follow the same sequence. This process is greatly facilitated if the work of the curriculum is arranged into units. As stated earlier, the units may be derived from the chapter headings of the course textbook, which makes the organization of course materials somewhat easier and takes a considerable burden off the shoulders of the teacher. However, other arrangements may be used as the teacher desires. For example, many teachers have constructed their own units for a course. Certainly teachers should feel free to organize the course around these units in an order or arrangement that they feel makes good sense. When we ask the students to choose a unit of work, we are in effect inviting them to select that chapter of the text that seems most interesting. After the group has made its selection, the students go to work with the text (or other material we have chosen) as the primary instructional resource in an attempt to learn as much as possible.

Changing Teacher Role

In a learning co-op, the teacher maintains the role of leadership in the classroom while the student begins to take more responsibility for learning. This is fundamental to a successful program. When survival groups are set to work on units of their choice, it should be made clear that students are not on their own; rather, the primary source of instruction has shifted from the teacher to many different sources, for example, textbooks, workbooks, manipulative devices, geoboards, audiotapes, filmstrips, and so on. We are trying to provide for variation in ways of learning by having ready for student use varieties of printed materials and other media. The student will move from one instructional resource to another as needed until the proper mix is found. Although teachers are no longer "up front" directing each step in the learning process, their role is in no way diminished. The students are offered the freedom to work, but the degree of freedom they have is a function of the amount of freedom and responsibility the teacher thinks they are ready to assume. For teachers to abandon the traditional role of "explainer" as their primary function is no great loss. Although there is a tendency to focus on the teaching skill of "explaining," being a successful teacher also involves many other skills. In addition to being skilled at organizing mathematics experiences to help students learn, teachers must also be skilled at making important human judgments. These judgmental skills include deciding how much responsibility a student can take and making adjustments in light of progress, determining how much support might be needed to enhance self-image, and deciding after diagnosis what kind of material or activity is best for student progress. These judgmental teaching skills traditionally take a back seat to the explaining skills. In a learning cooperative the relative importance of these two types of skills may be reversed, and the teacher's role altered correspondingly.

Helping Procedures

As teachers move out of the role of being the primary source of information, it becomes important for students to know exactly how they will proceed. The way the class starts each day is a good case in point.

Let us assume that each group, as a result of group discussion and consultation, has selected its unit of work. The students, on arriving in class, should form their groups and immediately begin working without any invitation from the teacher. In this way all available time can be used.

Experience has shown that one good way to initiate such behavior is for the teacher to come into the room and remain silent. When three or four minutes have passed and the students have not yet started to work, the teacher quiets the class and asks them what they are waiting for. In response, students generally ask, "Should we begin?" One way to reply is, "Why not? You know what you have to do. What are you waiting for?" With no more than that, students will often form their groups and go to work.

This exchange is usually repeated at the start of each class for several days. With one class, a teacher finally stood just outside the room but did not enter and could not be seen. He stood and listened. Several minutes after the period began, the students could be heard wondering out loud about the teacher's absence. Then a surprising thing happened: the students decided to begin working anyway; they formed their groups and went to work. Hearing this, the teacher entered the room to find virtually all the students busily engaged in their survival groups. From that day on, the students did not need the teacher to begin and usually began on their own with a minimum waste of time. A poster displayed prominently in the room served as a ready reminder of procedures (fig. 6.1).

Once everyone is working, the question arises as to what students are to do if they get stuck. When students run into difficulty, they should know exactly what steps should be followed to get the necessary help. The effi-

Fig. 6.1

cient use of the teacher's time can influence the eventual success of the program. It is generally good practice to look first to other group members for assistance, and last to the teacher. Thus, the teacher has jumped from being the first helping agent to being the very last resort. The intervening steps and resources should be clearly enumerated.

After trying fellow group members, students should seek an alternative textbook in the same area of study. They should turn to the index, look up the topic, locate it in the text, and then attempt to work with the alternative explanation offered there.

Should that fail to eliminate the problem, they may turn to an additional source. For example, the teacher may have made an audiotape of explanations and exercises in this area. Commercial tapes may be available. A programmed learning book may be helpful, or perhaps a filmstrip or film loop can offer some insights. Units involving manipulative devices may have been constructed by the teacher and prepared for just this problem. Whatever the medium, all available instructional materials should be carefully investigated and employed in an attempt to resolve any developing difficulties.

If the problems are still not resolved, other students from other groups may become sources of help, since they may already have completed the unit in question. And finally, all else failing, the teacher should be asked to render some immediate first aid.

A program is self-defeating if it results in students sitting around, wasting time waiting for help from the teacher. Carefully establishing this hierarchy for help is one way to insure that the teacher's time will be used in a manner that does not adversely affect the program. At the same time, the students are being assisted toward greater independence, an important step in the process of learning how to learn. A poster outlining these procedures for help reminds the students what to do (fig. 6.2).

It is important to make available to each student, if possible, the answers to whatever problems that student is working on. The proper use of answers should be discussed with students so that the answers are not abused. Experience has shown that once the students taste the satisfaction that results from solving problems first and checking after, the abuse of answers is diminished. In addition to checking their work, students may turn to

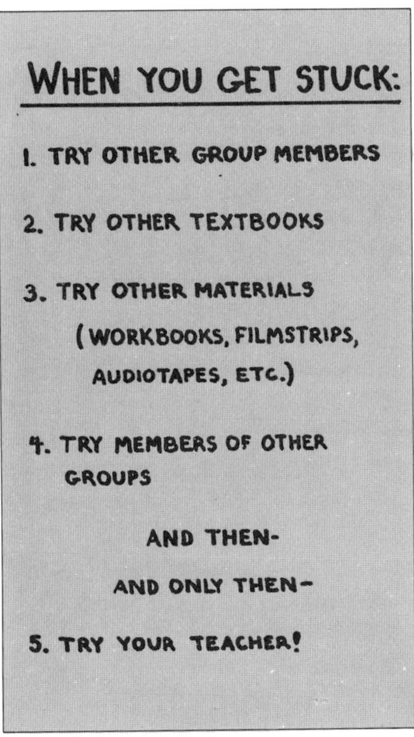

Fig. 6.2

answers if they are stuck. It is not unusual for students to look at an answer to a problem that they could not solve and then use the answer to construct a solution. The problem is thus exposed to view, and students are then able to find a solution. This is certainly a valued mathematical problem-solving technique, and one our students should know.

Testing

Testing should also be discussed with students. We are well aware of the importance of test results in arriving at students' grades, but in the learning co-op, testing is primarily used for determining the mastery of content. The way in which the tests are given is also important. In one sense traditional testing is unrealistic. Students have an opportunity to study something for an arbitrarily determined length of time and are then given a single chance to show if they have indeed learned. This situation is tantamount to asking someone to assemble a new toy after being allowed to read the instructions, which are then taken away. The toy is to be assembled from what is remembered. It seems that the significant factor in this situation is that they have only so much time, not whether or not they have learned. Why not allow the students to proceed as they would outside of school? The closest we come to this now is the open-book test, which most students dislike.

To keep the emphasis in testing on mastery, we ask individual students to determine for themselves whether or not they have learned the mathematics of a given unit. This is done by having them take the end-of-chapter tests found in almost every mathematics textbook. If the answers are available to them, the students have immediate feedback on how much they have learned. Should they feel they are ready, they then request a teacher-made test on the unit. The students do all this with the knowledge that if they should fail the teacher-made test—if they do not demonstrate mastery—there is no penalty. The test is used as a diagnostic instrument, and the students return to the unit for further study, to repeat the testing procedures once again when they feel they have indeed mastered the content.

For those students who do well on the test, the teacher records the grade, and they are ready to select a new unit. The only additional burden on the teacher is the need to construct more than one test on a given unit. This problem can be dispatched by making three virtually equivalent tests for each unit instead of one—simply alter the values where possible, being careful not to introduce unexpected difficulties in the process. Each test can be marked with some symbol so that the forms can be easily differentiated. For example, a group of three tests made for unit 6 could be marked, 6-1, 6-2, and 6-3. It is surprising how quickly students learn what this code means no matter how we try to identify the test forms. It is even more surprising that experience has shown that students generally will quickly let the teacher know if, by chance, they are given a test form that they took previously. It seems that once students know that they may take as many tests as necessary to establish mastery, most of them remarkably avoid any prior knowledge of the test questions. It sounds strange indeed, but remember we have changed the game, and in the process we may well change how students respond. This is what the learning cooperative is all about.

We must take care that in retesting, we are helping a student only to gain mastery, not more. Just as it sounds odd to teachers to let students retake tests, so does it sound equally odd to students. Since they want the highest possible mark, some students who do well want to retake tests to try for a perfect grade. This is a waste of time better spent on another unit. We are not trying to create specialists on, say, unit 4, linear equations.

Homework

How does homework fit into the learning co-op? The same pattern developed earlier is applied here. The students should discuss the importance of homework and consider reasons for doing it. Homework might be seen as—

1. a chance to practice material introduced in class;
2. a chance to make additional progress through the units;
3. a chance to prepare for the study of a new topic.

To be consistent with the sense of responsibility established in the survival groups, individual students may or may not decide to do homework on a given day as they or their group sees fit. This freedom must also be a product of student maturity and the teacher's honest feeling about the worth of the assignment to the youngster. "Busy work" will leave students with the same feeling we had when our teachers made us do it. It is true that we all have to do things we do not like because they will be good for us; it is also true that there is no conclusive evidence that homework is indeed good for us. What is more, the negative feelings engendered by busy work defeat our purposes. As the leader, the teacher may well decide that a particular student or group of students should be given a homework assignment. This is always the teacher's option.

The students should keep a careful record of assignments. The record will prove to be useful, especially in trying to determine the best course of action to help a floundering student.

Keeping Track

With all that is going on, the teacher may well wonder how to keep track of everything and everyone's progress. One nice thing about emphasizing student responsibility is that the only record the teacher needs to keep is the one that is generally kept, a record of examination results. And in this record only those tests on which the students demonstrated mastery are included. These scores may be recorded in the regular grade book. Teachers keep in tune with each student through a student progress record sheet that is distributed to each student for entries and then stored in a folder at the teacher's desk. The students should have free access to the folder whenever an entry has to be made. In figure 6.3 a sample form is shown partially filled in, but teachers may devise their own, providing space for all the items they feel are important. The student marks the sheet when a unit is chosen and started, when a test is taken, when a unit is completed, when group members change. If students keep the records up to date, the teacher will be able to sit down with the progress sheets perhaps once a week and at a glance see where each student is and how long she or he has been there. In these circumstances teachers can better arrange their plans. For example, in going through the record sheets at the end of the first semester, suppose we noticed a form (fig. 6.3) kept by a student in our algebra class. We might have several questions:

Which unit is Stuart on now?
Unit 2: Language of Algebra

When did he start it?
On 22 December. Just before winter holiday.

How long has he been on it?
Almost four weeks of school time.

Who is in his survival group?
Bob A., Charles W., Louis T., Leroy R.

Has he worked with any other students?
No.

The information from this progress record sheet indicates that Stuart and his survival group should be checked first thing tomorrow to determine why they have been on unit 2 for so long. Perhaps they are in trouble and require special help. Perhaps they are trying to attain 100 percent mastery before taking the test. More information is necessary, and so Stuart and the other members of the group must be seen.

The fact that the same students have remained in this survival group all year should also be explored. Why haven't they worked with other students? Although this is not necessarily a problem, it may be advisable to try to encourage them to mix with other students.

Student Progress Record					Name	Stuart Newcombe		
Unit	1. Sets	2. Language Of Algebra	3. Logical Thinking	4. Directed Numbers	5. Rational Numbers	6. Equations-Inequalities	7. Algebraic Problems	8. Graphs
Dates	10/1 – 10/22	12/22	10/23– 11/7	11/8 – 11/23			11/23– 12/8	12/9 – 12/21
Tests	10/17 10/22-78%		11/7-85%	11/21– 11/23-100%			12/8-76%	12/20- 12/21-78%
Survival Group	Bob A. Charles W. Louis T. Leroy R.	Bob A. Charles W. Louis T. Leroy R.	Bob A. Charles W. Louis T. Leroy R.	Bob A. Charles W. Louis T. Leroy R.			Bob A. Charles W. Louis T. Leroy R.	Bob A. Charles W. Louis T. Leroy R.
Unit	9. Algebraic Phrases	10. Factoring	11. Rational Phrases	12. Real No.-Radicals	13. Quadratics	14. Probability	15. Indirect Measures	
Dates								
Tests								
Survival Group								

Fig. 6.3

Using Student Information

The considerations above are a sample of how a teacher might react and the kinds of considerations that go into planning for class when a learning co-op is conducted. The emphasis in planning has changed from being on the arrangement of subject matter to giving specific attention to individual problems. Of course, teachers may also plan explanations of certain mathematical concepts that may be a problem for a student or group of students. Whatever is done, however, will be the result of trying to meet an identified need. The broad approach of traditional instruction is replaced with specific and particular help for individual difficulties, whether in mathematics or in personal areas. In addition to the information gained from the progress records, teachers, as they move about the room, will learn more about each student than was possible while standing at the front of the class. Observing students working in peer groups, becoming involved with their specific problems in mathematics, and offering assistance as the need is determined will leave teachers with insights about student abilities and personality that would be impossible to develop in traditional instruction, where students are confined to their seats and communication with others is discouraged.

This information helps to build a total picture of the youngster, giving teachers a better opportunity to offer what is needed.

The progress record sheets are kept current by the student but stored by the teacher to prevent loss. In addition to serving as a means of keeping track of all students, these records also present a quick picture of all the work accomplished by each student. In this way records become an important vehicle with which the students as well as the teacher can keep track of progress. Comparisons of current work with the student's earlier work may easily be made. The progress, or lack of it, can readily be seen.

Competition and Grading

Since only passing grades are recorded in the grade book, the basis for grading cannot be the comparison of one student with the others in the class. Indeed, if such competition is established, it may well undermine the successful operation of the learning co-op. Competition between students may hinder their cooperation, which is the basis of the learning co-op. The only competition desired is that of a youngster with himself. *Am I doing better?* This kind of competition is constructive. It may well lead to increased effort for the right reasons: not to do better than someone else, but to better oneself. There is a feeling that we must encourage competition in schools because our students live in a highly competitive world. Although it is true that society is sharply competitive, perhaps we shall enable our students to be better prepared to enter this competition if we encourage cooperation and build a strong and healthy self-image for each student. Who can say? In class, competition has a destructive character, since it results in winners and losers. Often a single winner prevails over a multitude of losers. Why not help each student become a winner?

Even though every school has its own grading policies, in a learning co-op it is best to grade individuals only against themselves. What is the youngster capable of and how much of this has been realized? Subjective judgment must necessarily be a part of the analysis. It is difficult at best to be completely objective in any situation. As long as teachers make judgments about what shall go into a test in the first place, a degree of subjectivity will always be present. They should, however, strive to be aware of their subjectivity and try to be as fair as possible in making these professional decisions.

Teaching Whole-Class Lessons

Does the use of survival groups imply that whole-class lessons will never be taught? Not at all. If teachers feel that the entire class would benefit from a more formal presentation in a particular area, they should by all means make such a presentation. They may also find, especially at the start, that some students need a teacher to "teach" them. Some students are more dependent than others. In the learning co-op teachers are free to use their time as they think best. They may make presentations to a group or groups of students to help them toward greater independence. They offer what they think *each* student needs to move ahead in mathematics and to grow and mature personally—no more, but no less.

Discipline

It may sound like a paradox, but it seems as easy or easier to maintain good discipline in a learning co-op than in the traditional classroom. For one thing, teachers are physically closer to the students as they move about the room, and their very presence exerts some control. Should a problem arise, they have the freedom to deal with it directly and tactfully, which minimizes complications. For another thing, group pressure is an important aid to the teacher in maintaining discipline. Disruptive students may interfere with the progress of their group and bring down the wrath of their peers. In addition, in the cooperative, students rarely can disrupt other students beyond their own group, but in the traditional class everyone becomes involved. Finally, although concern for maintaining order is intensified when students work on a variety of different topics, interesting work is perhaps the best deterrent to disruption. Students are often much more industrious about mathematics when they are permitted to select their units of work, as they do in the learning cooperative.

Summary

A learning cooperative has been described in which each student assumes responsibility for his or her learning and the teacher maintains responsibility for leadership. The students organize themselves with the teacher's help into survival groups for the study of the curriculum. Each group is free to select a unit of work from the given syllabus and to proceed to learn, using the textbook or other material. The teacher circulates about the room, providing help (i.e., *teaching*) as needed. With some students, teachers wait to be asked; with others, who may never ask, they must offer assistance. Teachers must continually make such decisions.

As the students work, they assign their own homework and determine their own readiness for tests. With extremely immature students, the teacher may well make decisions on when to give tests, too. On request, the teacher tests students the following day, using teacher-made tests. Multiple forms of tests are made for each unit as required. Should students demonstrate their mastery of a unit, their test score is entered in the teacher's grade book, and they go in to a new topic. If some students fail to master the content of the unit, the test is used for diagnostic purposes to indicate areas of further study. The students take another form of the test later to attempt to find out what they have learned. Should the materials of instruction, such as the textbook, fail to assist students, other materials are employed; the teacher offers help after all else has failed.

What becomes of a survival group if some students progress faster than others? It is not unusual to find that students will consider themselves a group despite the fact that each group member may be working on a different topic! Their ingenuity is boundless. They help each other when necessary, discuss problems as they arise, assist each other with assignments—in short, they exhibit all the basic characteristics of a survival group. Students are not held back by other group members, since the whole idea, as they see it, is to learn as much as they can in as little time as possible.

The special requirements of a learning cooperative in mathematics are minimal. No special classroom space is necessary, and the survival groups can function just as well in the so-called open school setting as in an old-fashioned desks-nailed-to-the-floor classroom. The curriculum is not special, the only change being the adding of more concepts if the syllabus of the course should be completed before the year's end. Students may then either continue vertically through the mathematics curriculum or select from optional topics for that course.

It is also possible for the learning cooperative to be used as a transitional stage by a mathematics department that is attempting to move from a traditional setting to a more open one. The gradual movement from a teacher-dominated class to a student-centered one may be one effective way of bringing about such a change. The important idea is that it is possible to provide for student differences in any setting without adding to the burdens of already overworked teachers and without isolating students from each other.

Despite major curriculum changes in the past two decades, relatively little has been done to alter the way in which mathematics is taught. The learning environment has remained fairly constant. It seems that moving students to the center of the learning process in mathematics classrooms is long overdue, as is a corresponding change in the role of the teacher.

Establishing a learning cooperative using survival groups is but one way to revise both teacher and student roles to the advantage of each. Teachers can make better use of all their professional skills. Students can learn to assume responsibility, to be considerate, and to acquire mathematical concepts and skills more effectively. The beginning of a lifetime process of learning may be started.

7

Organizing for Simulations

John C. Peterson

Role playing. Performance based. Human relations. Problem solving. Real life. Task oriented. Decision making. Confounding variables. Extraneous data. Inaccurate data. Insufficient data. Many formats. Small groups. Large groups. Community resources. Nontraditional materials.

Does this sound like a mathematics classroom? It can, if the classroom is organized around a simulation approach.

What Is a Simulation?

There are many definitions of simulation. In this essay, simulation is defined as realistic games or models that provide participants with lifelike problem-solving experiences. A simulation requires each participant to make decisions based on previous training and available information. After participants make a decision, they are provided with opportunities to see or discuss one or more possible consequences of this decision.

Although many games are available in mathematics, most do not qualify as simulations according to the definition established above, since they do not pertain to the real world. However, in the minds of many people the distinction between games and simulations is not clear. Some scholars apply the word *simulation* only to experiments involving computers; when human players make the decisions, the term *game* is used. For our purposes, a simulation will be considered a real-world laboratory lesson—a lesson based on concrete reality.

> While it is true that mathematics is an abstract system which children
> must learn to manipulate, the great virtue of that system is its relation to

113

concrete reality. Word problems don't really answer the need, for they are part of another abstract system, language. [Zuckerman and Horn 1970, p. 218]

Does Research Support Simulation?

Most simulations are designed to teach concepts of social studies, business, or economics rather than of mathematics. Although little research has been done on the effectiveness of simulations in any one area (particularly mathematics), two educators have examined the research that has been done.

Hunt (1972) listed five postulates about simulations that continually appear in the literature:

1. Simulation provides for student practice in communication skills by creating a controlled atmosphere in which communication can occur.
2. Simulation provides students with the opportunity to sharpen their problem-solving skills.
3. Simulation creates more student interest in learning than is created by more traditional teaching strategies.
4. Simulation provides a physical as well as a mental activity.
5. Simulation provides a laboratory situation where complex interpersonal behavior can be examined.

Chartier (1973) summarized research findings on simulation games and reached the following conclusions on their learning impact:

1. As a device for motivating students, simulation games seem to be exceptionally effective.
2. Students seem to achieve cognitive objectives by participating in simulation games as well as they would through other teaching-learning approaches, especially more conventional classroom activities.
3. Simulation games teach students the winning strategy.
4. Simulation games may influence a shift in student attitudes.
5. Simulation games seem to be effective tools for teaching students decision-making and problem-solving skills.
6. Simulation games have a way of reaching some students that other methods fail to accomplish.

The Goals of Simulations

Simulations can be used to meet educational goals as (1) a culminating experience, (2) a motivational device, or (3) a continuing core. A simulation can be used as a *culminating experience* for a unit to give students an opportunity to apply and develop further the mathematics they have just studied.

A simulation that could be used as a culminating activity follows. This example, "The Eyes Have It," includes some elementary probability and statistics; it could require the cooperation of art students. Students can be placed in groups of three or four for this simulation.

The Eyes Have It

Before a new product is put on the market, it goes through a period of testing to determine what people like or do not like about it and to determine whether they will buy it. People who conduct this testing are called market researchers. Market researchers not only test products but also try to find out what type of package designs will prompt the most sales.

Your group is a market research team for the What Not Company. The company's president thinks that sales of What Not's breakfast food, Gulp, may be increased if a new design for the Gulp box was made. The art department has submitted several designs. Your market research team is to determine which of these designs will result in the most sales. One way you can do this is to take a random sample of people's opinions about the new design.

How can you make sure that the sample is random? What questions should you ask? How many people should you sample? Will small children, teenagers, and adults pick the same design? How many people in your sample would actually buy Gulp? What age group will buy the most? What other questions should be considered?

On the basis of your group's random sample, which design would you recommend that the company use? How should you present your data to the president?

A simulation that is to be used as a *motivational device* comes at the beginning of a unit. Such a simulation can be used to help students see the need for studying more mathematics. An example of this type of simulation is illustrated in figure 7.1. The example requires students to verify an inventory form. In order to complete the verification, the student must be able to compute the number of units dispensed from bulk quantities. Since the units used for bulk quantities differ from the units used for dispensing, the student must either know or be able to determine the necessary conversions.

A conversion chart is given on another page of the simulation. However, if the teacher wanted students to set up their own conversion charts, the chart could be withheld from the simulation. The simulation would act as a motivator for making up a conversion chart and then for using the chart to complete the inventory form.

Simulations may be used as the *continuing core* of a mathematics course. This is most pertinent to vocationally oriented mathematics courses such

Hospital ward clerks should be able to compute the number of units dispensed out of bulk quantities of drugs and medical supplies so that, when necessary, they can verify the inventory records of the pharmacy department.

You have been asked to verify the record shown below prepared by the pharmacy department. There are several errors; correct them before signing the form. Notice that sodium bicarbonate is purchased in quantities of 5 kg and dispensed in quantities of 10 g. One kg contains 1000 g. Therefore, the following computation was made:

$$5 \times 1000 \text{ g} = 5000 \text{ g}$$
$$5000 \text{ g} \div 10 \text{ g} = 500 \text{ units}$$

A check is placed after the 500 in the "No. of Units" column to show that it is correct.

No. 523782

MOUNTAIN HOSPITAL
PHARMACEUTICAL
INVENTORY RECORD

Date _______________ 19 _____

Items	Bulk Quantity	Dispensing Quantity	No. of Units
Sodium bicarbonate	5 kg	10 g	500√
Quinine	1 l	5 cm³	2 000
Iodine	2 kg	2 g	1 000
Amphyl	4 kg	100 mg	40 000
Staphene	5 l	250 cm³	200
Saponated Cresol	1 l	4 cm³	250
Alcohol	5 l	100 cm³	50
Penicillin	1 l	20 cm³	500

Signature ________________________

Fig. 7.1. Adapted from Huffman et al. 1975 (p. 175)

as business, consumer, and shop mathematics. In some of these courses, one simulation or a series of related simulations might run for the entire course. In the business mathematics course, the activities of an office and the problems, both mathematical and nonmathematical, that arise during the simulation can provide the springboard for all formal instruction. A simulation that could be used at the beginning of a business mathematics course, "The Coach's Dilemma," is given at the end of this essay. This simulation is used to help motivate students to strengthen their weak areas.

What Simulations Are Available?

Since most simulations are in other academic areas, where can one find simulations that are useful in teaching mathematics? Zuckerman and Horn (1970; 1973) include reviews of professionally produced simulations and games. A description of the simulation "Supermarket" is given in figure 7.2. In mathematics, Zuckerman and Horn (1973) include reviews of thirty-five simulations or games that are appropriate for elementary school, thirty-six for junior high school, and twenty-four for high school. This sounds like an extensive list, but it includes nonsimulation games.

Many simulations can be located through Educational Resources Information Center (ERIC) listings by looking for materials that were developed for career education. For example, the two Taoscore citations listed in the References contain many elementary school simulations that include mathematical concepts among their objectives. Although a title does not provide an adequate description of the simulation, Taoscore has simulation titles at the first grade level such as "Bank Simulation," which includes adding and subtracting to ten, writing numbers and figures to ten, making change with play money, and counting money up to two dollars, and "Popsicle Sales Simulation," which includes the mathematical concepts of counting money, making change, and adding prices. During the second grade, students may participate in the "Egg Hatchery Simulation" (reading degrees on a thermometer and counting days of incubation on a calendar) and the "Plant Nursery Simulation" (making change, setting prices according to costs, and keeping records). In the fourth grade there is the "Easter Egg Manufacturing Simulation" (adding the cost of production, determining the price of a product, adding sales proceeds, and subtracting costs from proceeds to determine the amount of profit) and the "Restaurant Simulation" (comparative shopping, planning amounts of food needed, measuring ingredients, and cashiering). Fifth graders may pursue "Sewing and Metal Shop Simulation" (measuring in inches, computing costs and profits, and a review of annual finances of class projects), or "Mass Communications Simulation" (timing programs in seconds and minutes). These are just some of the simulations available from Taoscore.

Supermarket

Creative Studies, Inc.   c. 1969

Playing Data

Age Level: • Junior High
                  • Special Education

Special Prerequisites: Knowledge of addition, subtraction,
                          multiplication

Number of Players: 8 to 40 divided into 4 to 8 teams

Playing Time: 1 to 4 hours in 45 minute periods

Preparation Time: 1 hour the first time only

Materials: • Board
                    • Players' manuals
                    • Administrator's manual

Comment

• Interactive, with elements of both competition and cooperation
• Probabilistic: minor random events influence play
• Free form role play within assigned role positions
• Quantitative outcomes
• Individual and team play
• Play involves skill and decision making   (D.Z.)

Summary Description

Roles: • Some players represent store managers
              • The rest represent buyers

Objectives: • The store managers try to maximize sales
              • The buyers' objectives are to maximize purchase value
                within a given budget

Decisions: • The buyers must decide what items to purchase and
                which store offers the best bargain on the item they
                wish to purchase. The managers must decide on which
                items to increase price and which to decrease price

Purposes: • Quantitative skills in supermarket buying and selling
              • Multiplying one and two digit numbers
              • Finding and comparing percentage

Cost: Unknown

Producer: Not published but can be played by contacting
            Creative Studies, Inc.,
            167 Corey Road, Boston, MA 02146

Fig. 7.2

Junior high school simulations are included in the BO-CEC materials
(Huffman et al. 1975). High school materials can be adapted from *Geometry: Career Related Units* (Pierro et al. 1973) or *Career Related Math*

Units (Pierro et al. 1971). A description of these high school materials is also available in Helling (1976).

Organizing the Classroom for Simulations

Using simulations may necessitate some changes in the organization of the mathematics classroom. Following are ten organizational considerations involved in using simulations. All of these should be reviewed as a teacher prepares a simulation.

<table>
<tr><td colspan="2" align="center">Organizing the Classroom for Simulations</td></tr>
<tr><td>1.</td><td>New role functions are assumed by the teacher.</td></tr>
<tr><td>2.</td><td>New role functions are assumed by the student.</td></tr>
<tr><td>3.</td><td>A task-oriented atmosphere prevails in the classroom.</td></tr>
<tr><td>4.</td><td>Human relations and interpersonal skills are interwoven with mathematical learning.</td></tr>
<tr><td>5.</td><td>Sometimes students work alone; sometimes they work in small groups; sometimes they work in large groups—variety prevails!</td></tr>
<tr><td>6.</td><td>The room looks different.</td></tr>
<tr><td>7.</td><td>The variety of instructional materials and equipment will increase.</td></tr>
<tr><td>8.</td><td>The evaluation of students relies heavily on performance-based and time-based testing techniques.</td></tr>
<tr><td>9.</td><td>Some simulation activities may take place outside the classroom, and appropriate school personnel must be alerted to this fact.</td></tr>
<tr><td>10.</td><td>Parental cooperation is sometimes required for simulation activities.</td></tr>
</table>

1. Developing new role functions for the teacher

Simulations give teachers a chance to assume nonauthoritarian roles. In one simulation, the teacher may take the role of an employer; in others, the personnel manager, statistical consultant, building contractor, home buyer, or consumer. If a simulation is well designed, the teacher can often stay in the background. Learning will take place because of the careful planning that the teacher has put into preparing the simulation.

2. Developing new role functions for the student

In one simulation a student might be a market researcher; in another, a supermarket manager; in a third, a bank teller. With each simulation comes the chance to assume a new role and to use and learn mathematics in that role.

3. Working in a task-oriented atmosphere

Each simulation requires the completion of some task. Instead of isolated mathematical problems that have little meaning in and of themselves, mathematical problem solving becomes a means to find answers to an interrelated and interlocking set of problems.

4. Getting along with others

In many jobs, people must work together, and they must get along with each other. A simulation often requires students to work with each other or with people in the community. If students are to complete their simulation successfully, they will have an opportunity to develop their skills in interpersonal and human relations.

5. Working in a variety of ways

Some simulations require students to work alone; the student is given a task and is solely responsible for its completion. Others, such as "The Eyes Have It," have students work together in groups of three or four. In small-group simulations, each student might perform the same task, and the group would pool the results. Finally, some simulations have students work in large groups. In a large-group simulation, each student may have a different task. The group cannot complete the simulation until all have finished their assigned tasks.

6. Rearranging the room

One time the room may be arranged so that students having the same role are grouped together. At other times, the classroom may take on the appearance of an accountant's office, an engineer's office, a salesroom, and so on.

7. Locating and using new instructional materials

Some materials, such as catalogs, maps, and business forms, can be brought into the classroom. Other materials cannot be brought into the classroom, and students must go to the source. For example, some information can be obtained only from certain government offices, such as the mayor's office or the city planner's office. Other information, which may not be available in the school library, may be available in the city library. Knowing how to locate new material and then being able to use and evaluate it is an important learning experience students get from a simulation.

8. Using performance-based objectives in evaluation

When students begin working, their performance will often be evaluated on more than getting the correct answer. Employers consider the promptness with which the task was completed. How efficiently did the employee work? Was the employee able to work effectively with others? Did the employee work independently or did each step have to be approved? All these factors can enter into how employees are evaluated. Whenever it is appropriate, evaluations similar to those used by employers should be used in a simulation.

9. *Arranging for activities outside the classroom*

In "The Eyes Have It," students conduct random samplings that include small children, teenagers, and adults. If the sampling is conducted during school hours, then school administrators should be informed. Another simulation might involve students working in the halls, cafeteria, library, or other school areas. School personnel in these areas should be informed of the simulation and the type of activity that students will be conducting.

10. *Obtaining parental cooperation*

Be prepared to answer parents' questions about the unusual way that mathematics is being taught. Good simulations have mathematical objectives and functional objectives. Show parents that the students not only are learning mathematics but also are learning how mathematics can be used in real life. Solicit parental help in designing other simulations.

Guidelines for Selecting, Adapting, or Designing Simulations

Simulations have been used extensively in economics, business, and social studies. Mathematics teachers can select from the few that are available in their subject, adapt simulations that were made for other subject areas, or design their own. Following are thirteen guidelines that might be used in selecting, adapting, or designing simulations.

Guidelines for Selecting, Adapting, or Designing Simulations

1. Decide on the objectives.
2. Prepare simulations that provide a true representation of the mathematical concepts or ideas being presented.
3. Promote student initiative and self-direction.
4. Create an atmosphere in which students can understand that external events often affect outcomes.
5. Introduce confounding variables.
6. Consider providing extraneous, inaccurate, and insufficient data.
7. Include one or more opportunities for making a decision.

> 8. Look for ways to promote the seeking of information from sources outside the school.
> 9. Use realistic source documents.
> 10. Look to the real world for simulation ideas.
> 11. Provide opportunities to assume roles.
> 12. Make sure that not all roles are employer/employee roles.
> 13. Develop a game type of format for some simulations.

1. *Decide on the objectives.* What mathematics will the students need for the simulation? Do they have the necessary background information? What should they be able to do as a result of completing the simulation? How can it be determined if they have met these objectives? What non-mathematical objectives are in the simulation? Were affective as well as cognitive objectives included?

2. *Prepare simulations that provide a true representation of the mathematical concepts or ideas being presented.* Since a simulation is intended to provide real-life representations and uses of mathematical principles, it is important that the materials be mathematically appropriate. If expertise in an area to be simulated is lacking, consult with someone whose vocation or avocation is depicted by the simulation.

3. *Promote student initiative and self-direction by using simulations that encourage students to use their own problem-solving routes rather than predetermined ones.* Teachers need to be receptive to the student's views on ways to approach a problem. Often a student will find a new approach that provides an alternative solution. Although not all will be more efficient than the "accepted" solution, some will.

4. *Create an atmosphere in which students can understand that external events often affect outcomes.* Students are aware of "chance" cards from games such as Monopoly. A well-publicized simulation in family living includes an occasional spin of the wheel of misfortune in order to expose students to the unexpected occurrences of real life. Try to provide for introducing the unexpected events that often alter a person's plans. These events should cause students to consider changing their mathematical treatment or problem-solving approach in the simulation. However, factors of chance should be proportionate to their presence in reality.

5. *Introduce confounding variables into simulation activities.* Most realistic problem-solving situations do not lend themselves to simple solutions. Although early simulations should tend to be simplistic, they should include more variables as the students progress. Not all these variables need to be of major importance, but they should be important enough for a student to have to account for them in order to solve the problem satisfactorily. How-

ever, a simulation should never be so complex that students cannot select and apply the necessary mathematical principles.

6. *Consider providing extraneous, inaccurate, inconsistent, and insufficient data.* People are seldom given the exact amount of correct information needed to solve a problem. Design simulations so that students can learn to select only the information that is needed to solve the problem and to ignore information that will not be used. Refrain from providing all the information that will be needed. Let the students decide what they need and where or how they will obtain the information that is missing. Then encourage them to verify that the information is correct.

7. *Include one or more opportunities for making a decision in the simulation.* As simulations get more complex, students should be provided with opportunities to decide what information is relevant to the problem's solution, such as what information is missing, what variables should be considered, what problem-solving approach should be used, and what mathematics is needed.

8. *Look for ways to promote the seeking of information from sources outside the school.* Design simulations so that students need to consult sources such as newspapers, business enterprises, government agencies, community institutions, banks, stores, restaurants, gas stations, builders, parents, and so forth. Have them check to see if their data are consistent. Do all sources agree? If not, what could account for the discrepancies or variations?

9. *Whenever possible use such realistic source documents as catalogs, business forms, maps, and blueprints.* Different companies, businesses, and government agencies have forms and catalogs that are unique to their operations. Obtain copies whenever possible, and if current catalogs cannot be obtained, attempt to procure the preceding catalog or the current one when it is replaced. Students will gain a better perspective of what is needed when they have to locate information in a 900-page catalog instead of from a single sheet distributed by the teacher.

10. *Look to the real world for simulation ideas.* Where and how do people use mathematics when they are not in school? How will students use their arithmetic, algebra, geometry, and linear equations? Get ideas from professional journals and research projects. Sources such as Johnson (1974), Pierro et al. (1973), *Man-made World* (Engineering Concepts Curriculum Project 1971), and *Statistics by Example* (Mosteller et al. 1973[a]; 1973[b]; 1973[c]; 1973[d]) have many good ideas. Talk to people who use mathematics, and find out how they use it. Then design simulations that illustrate these uses.

11. *Provide opportunities for students to assume roles.* To some extent,

students took the roles of a market researcher in the simulation "The Eyes Have It" presented earlier. What other roles can they assume—butcher, baker, computer programmer, engineer? Some simulations can involve several people assuming different roles. Role descriptions should be complete. What decisions are open to each player? Will each player reach the mathematics objectives?

12. *Make sure that not all roles are employer/employee roles.* People use mathematics in leisure and home activities. What mathematics is involved in planning an addition to a house?

13. *Develop a game type of format for some simulations.* A simulation game must encourage students to apply knowledge. In order to succeed in the game, students need to apply relevant information, skills, and insights to the game situation. The game type of format can provide students with prompt feedback on the consequences of their actions.

This essay has introduced simulations into the mathematics classroom, given some background on how to organize the classroom when using simulations, and provided some guidelines for selecting, adapting, or designing simulations. A lengthy simulation, "The Coach's Dilemma," follows. In this simulation, students are given a budgetary problem to solve. In order to solve the problem, they must work with such concepts as insufficient data, relevant data, reliable data, and extraneous data. This simulation is written for students. As you, the teacher, read it, reflect on the points that have been made on organizing the classroom for simulations. Use the following questions to determine how *your* classroom organization and teaching would change if you used this simulation.

1. What role will you play? The coach? Your own role as the classroom teacher?

2. What role will the students play? Accountant? Secretary? Administrative assistant? Student assistant? What authority will the student have?

3. Is this a task-oriented activity? What differences do you see between using this activity and using the classroom mathematics textbook?

4. Does this simulation provide opportunities for students to interact with others? What kinds of human-relations skills will students need in order to complete this simulation?

5. Does this simulation require students to work alone, in small groups, or in large groups?

6. What will the physical appearance of your classroom be as students work on this simulation?

7. What materials and equipment do you need to gather? What materials and equipment will students need to get?

8. How will you evaluate the students? Will they have to meet a deadline? Are you going to use criteria in addition to the "correct answer" for evaluating students? Will neatness count? Promptness? Efficiency?

9. Will students need to leave the classroom in order to complete this simulation? If so, what school personnel need to be notified?

10. What information, if any, needs to be conveyed to the parents?

11. Which of the thirteen "Guidelines for Selecting, Adapting, or Designing Simulations" were followed in writing "The Coach's Dilemma"?

The Coach's Dilemma

Each year the coaches of the Ames High School athletic teams must submit budget requests for the next year. The track coach has asked for your help with the equipment and travel portions of the budget for the next school year. On the way out the door, the coach remarks that travel costs will be 6 percent higher this year because of inflation. (This year's budget and next year's track schedule are shown in figs. 7.3 and 7.4.)

I. Salaries

Coach: 6 months @ $125 a month	$ 750.00	
Asst. coach: 6 months @ $60 a month	360.00	
		$1110.00

II. Equipment

24 sleeveless track jerseys (Style 18, Champion) @ $2.20	$ 52.80	
12 pairs boys' shoes @ $26.00 a pair	312.00	
1 discus, wood center, high school, men	14.65	
1 discus, wood center, women	13.50	
		$ 402.95

III. Contractual Services

Officials: 8 officials @ $25 a meet for 6 meets	$1200.00	
		$1200.00

IV. Travel

Fort Dodge Meet: 99 km @ $0.50 a km	$ 49.50	
Des Moines Meet: 45 km @ $0.50 a km	22.50	
Newton Meet: 82 km @ $0.50 a km	41.00	
Mason City Meet: 157 km @ $0.50 a km	78.50	
		$ 191.50
TOTAL		$2904.45

Fig. 7.3. This year's track budget

2 April	Waterloo	(Away)
9 April	Des Moines	(Away)
16 April	Fort Dodge	(Home)
23 April	Charles City	(Away)
30 April	Mason City	(Home)
7 May	Newton	(Home)

Fig. 7.4. Next year's track schedule for Ames High School

1. What information will you need to determine the travel costs?
2. Did the coach give you enough information to figure the travel costs? If not, how and where are you going to get the information?
3. Has the coach given you any travel information you do not need?
4. Can this year's travel budget be used as a reliable source of data?

It is not very likely that you can prepare a travel budget until you know how far the track team will travel to each meet, the number of "away" meets, and the cost for each kilometer of travel. Collecting relevant information is an essential step in solving many business problems. *Relevant data* are facts and figures that help you solve a problem. There is more to solving business-related problems than manipulating numbers. The correct numbers to manipulate must first be determined.

> Suppose the coach had merely shown you this year's budget and asked you to add 6 percent to this year's travel figure in order to arrive at the amount for travel in next year's budget. Would that procedure be a good one? What relevant data would help you make more accurate calculations for next year's travel budget?

It was determined that certain relevant data are needed in order to figure how much money will be needed for the travel portion of the budget. Did the coach's information include where the "away" track meets will be held? How many kilometers of travel will be involved in each "away" game? The cost for each kilometer of travel? The answers to two of these questions can be found in the information the coach provided. By looking at next year's track schedule, you can determine where the "away" meets will be held. The coach has told you that inflation will cause next year's travel cost to increase by 6 percent over this year's budget. The budget will tell you that the rate used in figuring travel costs for this year was $0.50 a kilometer.

However, none of the information that the coach has provided will tell you the distance between Ames and the cities where the track meets will be held. Thus the data to figure the travel costs are insufficient. You have *insufficient data* when relevant facts and figures are missing.

Many problems cannot be solved because certain necessary data are lacking. How will you find the distance between Ames and the sites of the track meets? What will be your source of information? A person? A piece of paper? This classroom contains some very good sources of data. Some of the sources you will find are catalogs, rate tables, maps, and distance charts. Many problems in this course will contain insufficient data, and much of the data you will need to solve problems can be found by using these sources. Once you begin solving problems, you will quickly learn that when relevant information is provided, it is often "buried" in a document or other piece of paper. This is just the way some problems are. People must learn that data and figures are not always presented just as they are in some of the mathematics textbooks you may have used. Data you need may be scattered among several documents, and sometimes you have to go searching for it.

> For instance, suppose you have just sold a customer a sweater that costs $15. What would be the amount of sales tax on this purchase? Do you have sufficient data?

Sometimes you are given data you do not need in order to solve a problem. Facts and figures not relevant to a problem are *extraneous data*. It is your task to sort out extraneous data. You really have no use for the distance between Ames and the cities that will be involved in "home" track meets, since no travel costs will be involved. What other extraneous data were you given?

Once you have the data you need to solve a problem, can you depend on the reliability of the data? *Reliable data* are error-free relevant data. Sometimes data that have no errors in mathematical calculations may contain other kinds of errors. Look at this year's budget (fig. 7.3). You cannot depend on the data that are given on the cost of travel to the sites of the track meets. The multiplication of kilometers times rate is correct, but the distance is wrong. The distance given is for only one way and does not include the total distance for a round trip. Remember— certain pieces of information in some sources would help you locate this type of error.

Locating errors is an important aspect of determining whether data are reliable. Here are some questions that may help you determine if data are reliable.

1. Is this the best source for the needed data? Is there a more up-to-date source?
2. Is there any way to determine how the data were gathered?
3. Is there any way to determine who gathered the data?
4. Are there any apparent mathematical errors?
5. Are you interpreting the data accurately?

Now that you have all the appropriate data, you can begin to calculate next year's travel budget.

1. In order to determine what it will cost to travel to Des Moines next year, you perform the following calculations:

$$\begin{array}{cc}
\$22.50 & \$22.50 \\
\times\ .06 & +1.35 \\
\hline
1.3500 & \$23.85
\end{array}$$

You decide that it will cost $23.85. Are your calculations correct?

2. Next you calculate the amount needed to travel to Charles City and to Waterloo.

$$\text{Cost each kilometer this year} = \$0.50 \qquad \begin{array}{c} 6\%\ \text{increase} = 0.50 \\ \times\ 0.06 \\ \hline .0300 \end{array}$$

$$\begin{array}{r} \text{Cost each kilometer next year} = \$0.50 \\ + .03 \\ \hline \$0.53 \end{array}$$

From the distance chart on a map (see fig. 7.5), you can find that the one-way distance from Ames to Charles City is 195 kilometers.

KILOMETERS CHART	Ames	Burlington	Cedar Rapids	Charles City	Clinton	Council Bluffs	Davenport	Des Moines	Dubuque	Fort Dodge	Grinnell	Iowa City	Keokuk	Le Mars	Marshalltown	Mason City	Newton	Oskaloosa	Ottumwa	Sioux City	Storm Lake	Waterloo
Ames		298	162	195	294	256	269	45	280	99	104	189	317	282	61	157	82	139	179	270	194	152
Burlington	298		165	347	184	437	129	254	240	393	213	123	72	574	259	394	227	158	118	550	486	261
Cedar Rapids	162	165		182	133	387	123	178	118	261	102	43	189	443	106	229	131	141	157	436	360	104
Charles City	195	347	182		318	431	308	215	220	194	181	228	376	330	139	47	189	232	272	369	258	81
Clinton	294	184	133	318		526	62	311	99	398	238	141	249	583	241	365	269	258	259	570	494	238
Council Bluffs	256	437	387	431	526		473	215	510	246	293	394	462	196	292	398	262	287	322	157	198	384
Davenport	269	129	123	308	62	473		258	112	373	188	89	194	557	222	355	217	203	204	544	468	228
Des Moines	45	254	178	215	311	215	258		296	141	78	178	277	324	76	202	47	97	138	300	235	168
Dubuque	280	240	118	220	99	510	112	296		319	224	138	285	466	227	266	253	258	274	512	416	144
Fort Dodge	99	393	261	194	398	246	373	141	319		206	292	418	186	162	154	183	238	279	193	97	175
Grinnell	104	213	102	181	238	293	188	78	224	206		105	235	390	49	207	31	55	96	374	301	110
Iowa City	189	123	43	228	141	394	89	178	138	292	105		147	476	138	275	134	120	136	462	387	147
Keokuk	317	72	189	376	249	462	194	277	285	418	235	147		601	284	423	249	180	139	577	512	295
Le Mars	282	574	443	330	583	196	557	324	466	186	390	476	601		347	280	368	421	462	39	89	350
Marshalltown	61	259	106	139	241	292	222	76	227	162	49	138	284	347		159	50	104	144	334	258	92
Mason City	157	394	229	47	365	398	355	202	266	154	207	275	423	280	159		191	249	290	319	217	128
Newton	82	227	131	189	269	262	217	47	253	183	31	134	249	368	50	191		70	110	343	279	141
Oskaloosa	139	158	141	232	258	287	203	97	258	238	55	120	180	421	104	249	70		40	397	332	160
Ottumwa	179	118	157	272	259	322	204	138	274	279	96	136	139	462	144	290	110	40		437	373	201
Sioux City	270	550	436	369	570	157	544	300	512	193	374	462	577	39	334	319	343	397	437		117	368
Storm Lake	194	486	360	258	494	198	468	235	416	97	301	387	512	89	258	217	279	332	373	117		272
Waterloo	152	261	104	81	238	384	228	168	144	175	110	147	295	350	92	128	141	160	201	368	272	

Fig. 7.5. Distance chart

$195 \times 2 = 390$ kilometers round trip

$$\begin{array}{r} 390 \\ \times \ \$.53 \quad \text{a kilometer} \\ \hline 1170 \\ 1950 \\ \hline \$206.70 \end{array}$$

The one-way distance from Ames to Waterloo is 152 kilometers.

$152 \times 2 = 304$ kilometers round trip

$$\begin{array}{r} 304 \\ \times \ \$.53 \quad \text{a kilometer} \\ \hline 912 \\ 1520 \\ \hline \$161.12 \end{array}$$

You conclude that next year it will cost $206.70 to travel to Charles City and back and $161.12 for the round trip to the track meet at Waterloo. Are your calculations correct?

Let's examine your calculations in the two questions above. Your calculations in question 1 are correct, but your answer is wrong. Why? Some of the data you used to solve this problem were not reliable. This year's travel budget was incorrectly calculated. It is true that 45 kilometers × $0.50 a kilometer = $22.50. What is the error?

The person who calculated this year's budget forgot that the track team makes a round trip for each "away" meet. Thus, the total distance for the Des Moines track meet should have been 90 kilometers instead of 45 kilometers, and the amount budgeted for the trip should have been $45.00 ihstead of $22.50.

Now check your calculations in question 2. Since the cost of travel for each kilometer in this year's budget is $0.50, an increase of 6 percent would represent an increase of $0.03, just as you calculated. Next year's travel will, therefore, cost $0.53 a kilometer.

3. Now calculate the total travel budget for next year. This year the total was $191.50. But in checking your calculations in question 1, you discovered that whoever figured this year's budget calculated the travel expenses using one-way trips instead of round trips. Because of this, the travel expenses were just half as much as they should have been. This year's total travel budget should have been $383:

$$\begin{array}{r} \$191.50 \\ \times\,2 \\ \hline \$383.00 \end{array}$$

A 6 percent increase would make next year's total travel budget $405.98:

$$\begin{array}{r} \$383 \\ \times\,.06 \\ \hline \$22.98 \end{array} \qquad \begin{array}{r} \$383.00 \\ +\,22.98 \\ \hline \$405.98 \end{array}$$

Exercises

1. Did you calculate the total budget for next year's travel correctly? Were you given any extraneous data? Were you given reliable data? Were the data insufficient? If $405.98 is not the correct total for next year's travel budget, determine the correct total.

2. Suppose that Ames High School decided to schedule a track meet at Dubuque. What are two sources where you could find relevant data for determining the travel costs for this track meet? What would the travel cost be for Ames if it had a track meet in Dubuque?

3. What should you budget for next year's equipment? Do you have insufficient data? Do you have any extraneous data? Are all your data reliable?

4. The coach tells you that you will need to budget for fifty sleeveless track jerseys, style 18. How much should be the total amount budgeted for these jerseys? Do you have sufficient data? Extraneous data? Reliable data? What sources do you need?

5. You should allow for twelve pairs of #501-Adidas "Tokyo 64" track shoes. How much should you budget for this item? Do you have sufficient data? Extraneous data? Reliable data? What sources do you need?

REFERENCES

Chartier, Myron R. *Simulation Games as Learning Devices: A Summary of Empirical Findings and Their Implications for the Utilization of Games in Instruction.* Paper prepared for Workshop on Simulation Games, American Baptist Seminary of the West and Holy Names College, fall 1973. (ERIC no. ED 101 384.)

Engineering Concepts Curriculum Project. *The Man-made World.* New York; McGraw-Hill Book Co., 1971.

Helling, Cliff. $\frac{"CD}{HD} = 1$." *School Science and Mathematics* 76 (April 1976): 313–20.

Huffman, Harry, Rick Mecagni, Celestine Mongo, and Clyde Welter. *BO-CEC Math Resource Guide.* U.S. Office of Education no. OEC-0-73-5230. Reston, Va.: National Business Education Association, 1975.

Hunt, Gary T. *Approaches to the Use of Simulation in Speech Courses.* Paper presented at the 58th Annual Meeting of the Speech Communication Association, Chicago, 27–30 December, 1972. (ERIC no. ED 072 500.)

Johnson, Donovan A. "Mathematics outside the Classroom." *School Science and Mathematics* 74 (February 1974): 129–34.

Mosteller, Frederick, William H. Kruskal, Richard F. Link, Richard S. Pieters, and Gerald R. Rising, eds. (a). *Statistics by Example: Detecting Patterns.* Reading, Mass.: Addison-Wesley Publishing Co., 1973.

———— (b). *Statistics by Example: Exploring Data.* Reading, Mass.: Addison-Wesley Publishing Co., 1973.

———— (c). *Statistics by Example: Finding Models.* Reading, Mass: Addison-Wesley Publishing Co., 1973.

———— (d). *Statistics by Example: Weighing Chances.* Reading, Mass.: Addison-Wesley Publishing Co., 1973.

Pierro, Mike, Claude Paradis, Richard Michalicek, and Arne Grangaard. *Career Related Math Units.* Minneapolis: Robbinsdale Independent School District 281, 1971.

Pierro, Mike, Claude Paradis, Doug Svihel, and Arnie Grangaard. *Geometry: Career Related Units.* Teacher's ed. Minneapolis: Robbinsdale Independent School District 281, 1973. (ERIC no. ED 085 548)

Taoscore Teacher's Guide: Phase 3 (a). Taos, N.Mex.: Taos Municipal Schools, n.d. (ERIC no. ED 096 414)

Taoscore Teacher's Guide: Phase 4 (b). Taos, N.Mex.: Taos Municipal Schools, n.d. (ERIC no. ED 096 415)

Zuckerman, David W., and Robert E. Horn. *The Guide to Simulation Games for Education and Training.* Cambridge, Mass.: Information Resources, 1970.

————. *The Guide to Simulations/Games for Education and Training.* Hicksville, N.Y.: Research Media, 1973.

8

Organizing for Mastery Learning: A Group-based Approach

Judith Harle Hector

*T*eachers at all levels encounter the problem of how to help the student who has not learned the mathematics expected for that student's age and background. At each level, the less skilled student usually receives some such label as "remedial," "developmental," "disadvantaged," "low achieving," or "slow learning." The slow-learning fifth grader who struggles with fractions can often be found some years later in a "developmental" mathematics class at a community college still essentially "mathematically illiterate" (Brown 1971, p. 124). One description of the process that produces the mathematically illiterate student seems particularly apt. Allendoerfer (1965) focuses on the lack of intuition as the key block to learning mathematics:

> The first step in the learning of any mathematical subject is the development of intuition. This must come before rules are stated or formal operations are introduced. The trouble is that we know all too little about how to foster the growth of mathematical intuition. It is here that children differ very widely in their rate of growth. The most "brilliant" see things in a flash; with the slower ones, we give up and have them memorize. When these slower students have memorized a number of bits of mathematics without intuition or understanding, they begin to get confused and their "computer jams." They may be able to add and multiply, but they do not know *when* to add and *when* to multiply. Before long they are mathematical failures, and no amount of remedial work seems to help very much. [p. 694]

Allendoerfer's pessimistic conclusion deserves to be challenged. Students who have been failures *can* succeed in learning mathematics when taught with a group-based approach to mastery learning. This approach has been implemented in a community college setting with minimal initial expense. Specific suggestions for implementation are given with references to school and college settings.

Carroll's (1963) paper entitled "A Model of School Learning" has stimulated interest in mastery learning. His model emphasizes five factors influencing learning. Students have variable amounts of *aptitude* (defined as the speed with which one learns), variable amounts of *ability* to understand instruction, and variable amounts of *willingness* to engage in learning. The school determines how much *time* students are allowed for learning and the quality of *instruction*. The time available to learn and the quality of instruction are two key factors in what has come to be known as mastery learning. As proposed by Bloom (1971) and Block (1971), mastery learning is a deceptively simple procedure. First, the best possible instruction on a topic is given. Second, the student is tested to determine any particular learning difficulties by pinpointing what has and has not been learned. Third, the student is retaught using alternative methods. This teach-test-reteach strategy is at the heart of mastery learning. The strategy is accompanied by a philosophy that "asserts that under appropriate instructional conditions virtually *all* students can and will learn well most of what they are taught" (Block 1973, p. 30).

Mastery Learning in Mathematics Instruction

A mastery-learning approach can assure more success in learning mathematics. A teacher who is considering whether to implement a mastery-learning model should first examine whether the model is appropriate to the students, the subject matter, and the school setting, and then how the model may be implemented. The following comments relate to such broader issues of mastery learning in mathematics instruction as flexible learning time, appropriate subject matter, and mastery level. The comments apply points made by Mehrens and Lehmann (1975) about mastery testing to mathematics instruction.

Flexible learning time

Different students learn at different rates. The mastery-learning model requires that the time available to learn a topic be variable. In the example to be discussed later, the date for testing on a specific topic is fixed. Faster students may take the test early; all students, however, are required to take the test by the assigned date. The test diagnoses what they still have to learn and indicates how much extra time they need to spend to master the topic. Sometimes they must devote several hours daily either to studying with the teacher or to studying alone. At the college level, the time of both teacher and student is more flexible than in elementary and secondary school. However, even at these levels it would seem to be possible to use group-based instruction for several days, with one day as a testing and "catch up" day. Students who have mastered the topic could then do enrichment activities while the teacher worked with those who needed more help.

Riedesel (1973) suggests that organizing a form of team teaching once or twice a week can provide the flexibility needed for reteaching. In a school having four classes at a particular grade level, one or more teachers could work with groups who had not achieved mastery while the other teachers made use of laboratory materials and experiments to give a greater depth of understanding to the topic just studied for those students who had mastered the essential skills. A teacher in a self-contained classroom can work with a group having varying abilities by discussing first the suggestions of students who have used the simplest approaches to a problem. The discussion then moves to the more complex approaches suggested by brighter students. To create some depth of learning, it is important to encourage multiple solutions to problems. An additional payoff is that the slower students arrive at a minimum level of mastery in the same time that the faster students explore several solutions. A mathematics corner for individual exploration can provide project cards, activities, supplemental worksheets, and puzzles for the faster students as a means of evening out the different amounts of time needed for mastery.

Another method of ensuring flexibility with time is by self-pacing using individualized programmed materials. Though much has been written on the successful use of this approach (Bittinger 1972; Burris and Schroeder 1972; Kulik, Kulik, and Carmichael 1974; Perry 1971; Spangler 1973; Wagner and Jones 1973), difficulties do exist in self-pacing with programmed materials. Those usually cited include the lack of independent study skills among students, the lack of adequate programmed materials at appropriate reading levels, and the tendency of students to procrastinate or progress too slowly through the materials (Spangler 1973).

Teachers who choose self-pacing through programmed materials need to ask whether they can handle thirty individuals studying different topics, how long the majority of students can sustain interest in mathematics with minimum discussion and exchange with their teacher and peers, whether students can function in a class with no more than an average of one or two minutes of direct attention from the teacher, and whether materials are available at every level of student ability. (Most of the May 1976 issue of the *Mathematics Teacher* focuses on self-pacing and individual and group approaches. The remainder of this essay discusses a group-based form of classroom organization.)

Appropriate subject matter

Some subject matter is more appropriate for a mastery-learning approach than other subject matter. The hierarchical nature of certain topics in mathematics fits a model in which one must master the basics before proceeding to higher concepts. For example, mastering computational skills with whole numbers precedes learning decimal computation. Mastery learning seems less appropriate in a "mathematics appreciation" course or in a course that covers units on set theory followed by topology followed by the use of linear equations. In that type of course, the mastery of the skills in one topic does not depend on the previous topic.

The subject matter should be important enough for mastery to be required. Students have a limited future in using mathematics if they do not learn to solve linear equations. Therefore, reteaching and retesting are appropriate. Yet, learning to prove that the square root of 2 is irrational may not justify extended study. At issue are the practical limits both to how long a student should struggle for mastery and to the teacher's skill in designing alternative methods of instruction. The teacher might present a standard proof in which the assumption that $\sqrt{2} = a/b$, where a/b is a rational number in simplest form, leads to a contradiction. A discussion of a similar proof that uses the fact that only integers can have rational square roots might follow for those students who had trouble with the first proof (Dolciani and Wooton 1970, pp. 464–65). A third proof, based on the rational-root theorem stating that $x^2 - 2 = 0$ has no rational roots, might also be helpful (Dolciani, Berman, and Wooton 1970, pp. 256–57). However, after these presentations, it may appear that a student is memorizing the proofs rather than reasoning logically. It may not then be possible to find meaningful alternative presentations to make reteaching and retesting possible.

Mastery level

The mastery level is arbitrarily set by the teacher. Testing determines if the minimum essential objectives have been attained. The less academically capable student who has continually scored lower in comparison to brighter students may, through retesting, attain the same score as the bright student on the test. Increased confidence and positive attitudes result. The bright student does not, however, receive the reward of being one of a few to score high. The goal is for *all* to achieve the objectives. Clever solutions and creative approaches to problems should be asked for and praised by the teacher, since they are not reflected in course grades in the mastery approach.

Course grades have functioned to rank students on a continuum from best to worst. The best received some reward from the high marks and the worst, some punishment from the low marks. The preponderance of A and B grades obtained in mastery-learning courses does away with this ranking information.

Well-defined set of skills

Mastery learning is appropriate for a well-defined set of skills such as computation and algebraic manipulation, in which it is possible to specify when someone has mastered the skills. Bloom (1971) points out that it is helpful if the topic is closed, that is, if it involves a finite set of ideas and behaviors on which teachers agree. Also, mastery learning fits subject matter involving convergent thinking, in which one can judge right answers, good solutions, and appropriate thought processes. Creative proof-making involves divergent thinking and thus fits a mastery-learning model less well. Similarly, the acquisition of problem-solving skills in the broad sense is more open-ended and hence less appropriate for mastery learning.

Tasks achievable by the majority

Finally, the learning tasks set up for mastery learning need to be attainable by the majority of students in the course. Certainly, the mathematical-literacy skills of computation and beginning algebra can be learned by the majority. And members of the Cambridge Conference on School Mathematics (1963) would even move "three years of top-level college training" in mathematics down into the secondary school (p. 7). The debate over whether that level of mathematical competency is within the reach of the majority will continue. Right now the decision regarding what subject matter is appropriate for mastery learning rests with the teacher. The labels attached to students, such as "remedial" and "slow learning," can influence teachers to expect too little of the student. Student success and the expectation of teachers that their students will master the topics taught often go together (Kipps 1970; Rummell 1958). The expectation of the teacher may be a factor influencing the success.

How to Implement Mastery Learning

Most mathematics teachers work in a group-based instructional setting. Frequently, the boundary conditions on an educational innovation require that it use no additional personnel and cost no more than the procedure it will replace. It has been possible to implement mastery learning while meeting these conditions at a community college in a developmental mathematics course covering basic computational and algebraic skills (Hector 1975) and at a large state university in an intermediate algebra course for students in liberal arts, business, and agriculture (Haver 1975). Several words of caution and suggestions for implementing mastery learning can be distilled from these experiences.

Words of caution

Mastery learning is not a panacea. It is important that administrators understand there will be no savings in cost or teacher loads. There have been mastery-learning situations in community colleges in which administrators have tried to combine three or more different mathematics courses having small enrollments into one section. One teacher was expected to deal with thirty to seventy students studying different topics in the different courses. The teacher became a mere grader of papers. Students struggled individually with programmed materials. Most discouraging of all, the fraction of students successfully completing the course fell below one-third.

A second note of caution involves paperwork. Anyone implementing a mastery-learning model must develop a system for writing alternative test forms, grading tests, and recording scores. One teacher counted the tests graded in a mathematics section of thirty students for one quarter—the number was 3000! A computer can lessen the avalanche of papers. Young, McKean, and Newman (1974) describe how they developed a large data bank of questions and answers stored for computer use. A computer program randomly selects questions and prints different tests with answer keys on accompanying sheets. In other systems, students take tests displayed at computer terminals. The computer selects the appropriate test based on previous test scores stored in its memory. The computer scores the test and records the score. Machine-scored tests, whether computer scored or scored by small and inexpensive scoring devices, save the teacher's time. In the absence of these technological aids, the teacher should consider organizing a notebook of tests and answer keys. It is helpful to have all tests, keys, and unit materials color coded. Different colors of paper can be used for different units. It may be helpful to ask students to use their own notebook paper for all work and answers to tests. The test sheet can then be reused, avoiding duplicating, stapling, and the filing of test forms. However, it is often faster for the teacher to grade tests if the student writes in the spaces provided on a test sheet. The teacher must consider the paper-handling job as the mastery-learning model is planned. The success or failure of mastery learning may depend on the efficient organization of paper handling.

Finally, the quality of instruction, whether teacher-based, programmed in written form, autotutorial, or audiovisual, must be such that students can learn. Simply retesting a student with no intervening learning becomes an exercise in frustration. In mastery learning, teachers or materials must be available so that several alternative approaches to a topic can be provided. There is more than one way to explain how a decimal point is placed in the quotient of two decimal numbers or how to approach a word problem. The challenge to teachers is to broaden their repertoire.

Organizational considerations

One teacher implemented mastery learning in a community college (Hector 1975). In this particular setting a teaching load was five sections, each consisting of ten to twenty students and each meeting three hours weekly. In addition to the fifteen hours of group instruction, the teacher was required to be available fifteen hours a week for office hours. Most of the reteaching and retesting was done in individual and small-group sessions during office hours. In public school, finding the time for the reteaching may be more of a problem. The public school teacher might consider organizing instruction so that students work together four days a week. On the fifth day, students who have mastered the weekly topic can do enrichment activities while the teacher works on bringing the rest of the class up to a mastery level. Parent volunteers can be used to supervise the enrichment activities while the teacher works with remediation. Peer tutoring is another way to do some reteaching. Still another alternative is for the teacher to decide which topics in the course are essential. Those topics can be taught in a mastery format interspersed among the less essential topics. Team teaching, asking for multiple solutions from more capable students, and providing activities in a mathematics corner for students not being retaught were mentioned earlier as ways of organizing to allow for differences in learning rates.

One helpful suggestion concerning organization is to begin small and expand slowly. For example, one university began with one section, developing materials and procedures before expanding to its present course enrollment of 600 students (Haver 1975). In an elementary setting, the teacher might want to experiment with one unit in a mastery format. A high school teacher might want to try mastery learning in one section or one unit before developing a system to handle all sections or an entire course.

Planning for instruction

How does one begin to implement mastery learning, given a teacher, students registered in classes, and the usual school setting? In the community college, the teacher divided the subject matter of the course, both text material and teacher-written material, into weekly units. Keeping in mind the question "What is it that I want students to understand and be able to do after they finish this unit?" the teacher keyed the unit's mastery test to the essential skills of the unit. This process is a part of good planning whether or not a mastery model is used. If the test is written after the unit has been taught, there is a tendency to postpone deciding what the objectives of the unit are until then. To illustrate, consider multiplication with fractions. Students should be able to represent with physical models the meaning of 3 × ¾, ⅔ × ¼, ⁴⁄₃ × ¼, and ¾ × 8. The first example, 3 × ¾, arises from making three batches of fudge, each using ¾ cup of sugar. Adding ¾ + ¾ + ¾ on a number line is an appropriate model. Giving someone ⅔ of ¼ of a candy bar can be seen in a rectangular diagram representing ⅔ × ¼ (see fig. 8.1). Students at community college often ask why the answer to ⅔ × ¼ is less than ¼. Their concept of multiplication includes the expectation that a product is always larger than the multiplicand. The use of physical models is therefore as important at the college level as in elementary or secondary school.

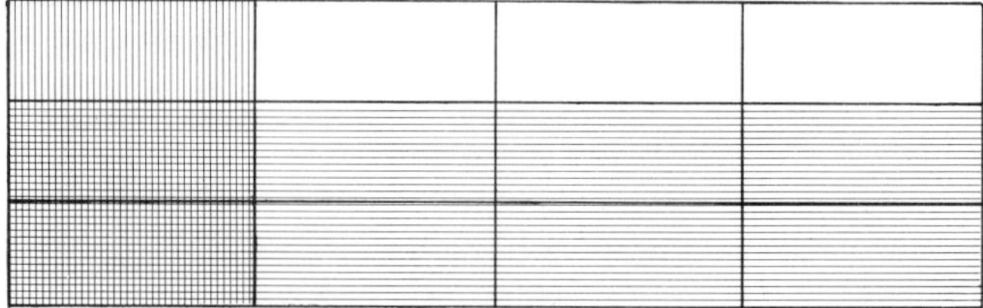

Fig. 8.1

The rectangular diagram used in figure 8.1 can also be extended to represent ⁴⁄₃ × ¼ by adding another row representing ⅓ of the rectangle to the top of the array. The last example of ¾ × 8 might arise in sharing eight candy bars with a friend so that one keeps ¾ of the 8. The process would include making four piles (two candy bars to a pile): one keeps three piles and gives away one. Besides attaching meaning to such problems as ⅔ × ¼, the students should be able to compute accurately when they encounter multiplication with fractions in textbook examples and problem situations outside the classroom. Computationally, the following examples differ:

⅓ × ½	2 × ¾
⅔ × ½	2 × 1¾
⅔ × ¾	⅓ × 2¾
2 × ⅓	1⅓ × 2¾

In addition to mastering the algorithm $a/b \times c/d = ac/bd$, the student will have to know how to factor, recognize a common factor, and remove a common factor to express several of the products above in lowest terms. The student will need to recognize that 2/1 equals 2 or be able to rewrite a mixed numeral as an improper fraction before using the algorithm on the last five of the examples above. The distributive property can be used to solve $\frac{1}{3} \times 2\frac{3}{4}$ as $\frac{1}{3}(2 + \frac{3}{4})$. Instructional strategies are planned around these concepts and skills.

The diagnostic nature of the tests and retests is an important aspect of mastery learning. The test should show what still needs to be learned before a retest is given. For example, a test item involving $2 \times 1\frac{3}{4}$ is often missed because the student has failed to use the distributive property. When the answer is given as $2\frac{3}{4}$, the teacher should stress in reteaching that both 1 and $\frac{3}{4}$ are multiplied by 2.

Commercial drill kits produced for elementary and secondary school students provide some ideas that have proved helpful in diagnosing computational difficulties. They offer a step-by-step progression from easier to harder computational problems on whole numbers, fractions, decimals, algebraic skills, and so on. Often, sections in programmed texts provide help in writing diagnostic test questions by detailing areas of frequent errors or misconceptions by students.

Pretesting

The preceding section discussed how unit tests can summarize how well a student has learned the unit. Some form of pretesting is needed to determine whether a student already knows the content of the unit. In the community college example, an initial pretest is given on the first day of class. Students who score well on arithmetic skills may transfer to a second course, which focuses on algebraic skills. (Some students choose to spend a school term reviewing arithmetic even though they score well because they want to gain confidence in their skills.) The completion of a unit indicates that a student is ready to begin a succeeding unit. The point of view is taken that test items are rough indicators of whether a student understands the area tested or whether more study is warranted. Self-paced instructional models generally have more detailed pretesting and posttesting for every unit.

Sometimes oral questions can elicit information for preassessment. Most students respond honestly to the question, "What is hardest for you in mathematics?" The teacher can then say, "Show me a problem that is confusing to you." The questioning technique can replace or supplement an initial pretest. Too often paper-and-pencil pretests are merely additional experiences of failure for slow learners.

Alternative teaching strategies

On the first day of class at the community college, students are given a written statement about the course. It explains that they will be graded solely on the basis of short weekly tests of ten to twenty minutes' duration.

The tests will be scored from 0 to 100 percent. The course grade will be an A for an average of 90 percent or above and a B for an average of 80 to 89 percent. The mastery approach is outlined with the statement that every student who scores below 80 percent is encouraged to get extra help and to take retests until a score of 80 percent or better is obtained. All are told that they can achieve a grade of A or B.

The teacher presents material, has students try examples in class, and walks from desk to desk, checking work and answering questions. Following the weekly test, the teacher encourages students to set up individual and small-group appointments if they receive low test scores. During their appointment, they receive additional help followed by a retest. Some of the more difficult skills have required as many as five retests before a level of 80 percent was obtained, but imagine the student's sense of success at that final achievement! (Imagine, too, how the teacher's techniques are refined, altered, and improved to reduce the number of retests required.)

One focus of the testing is on feedback to the teacher on what instructional improvement is needed. For example, the textbook used at the community college uses an equation approach to percentage word problems. The book says, let

"what" be X

"is" be $=$

"%" be times 0.01

"of" be times, and

numbers be themselves.

The translation process works for the three types of percentage problems that have the following forms:

What is _______% of _______?

_______ is what % of _______?

_______ is _______% of what?

Many students find this approach suitable and score well on test items related to percentage word problems.

The major difficulty with the equation approach, as revealed by the unit test, is that some students have great difficulty in matching a word problem with the equation form it contains. Therefore, the teacher needs another approach to the topic, such as the one suggested by Cole and Weissenfluh (1974). In this approach, students are able to relate percentage problems to their previous learning on proportion. They represent

What is 60% of 80? as $60/100 = X/80$

48 is what % of 80? as $X/100 = 48/80$

48 is 60% of what? as $60/100 = 48/X$.

The unit test can be given to determine who would benefit from learning another approach. However, in this instance, the teacher usually asks who would like some extra help. A small group is formed for this instruction; the rest of the class works on practice exercises. Most of the small group seem to grasp readily that the left members of the following proportions relate to the concept of percent as being some part of 100:

$$60/100 = X/80$$
$$X/100 = 48/80$$
$$60/100 = 48/X$$

The right member of the proportion is set up with reference to the concept of proportionality. For example, the students are given the following problem: A salesclerk tells you that the $48 calculator you want to buy is selling at 60% of its original price. What was the original price? The students find the left member of the proportion, 60/100, first. Initially, they may wonder whether the right member should be $X/48$ or $48/X$. However, they should reason that the $48 is less than the original price. They should think that since 60 is smaller than 100, so 48 is small than X, and they should choose $48/X$. The students are then tested. Individuals who still need help may go over the equation and proportion approaches for greater clarity or may be introduced to another approach, such as unitary analysis.

It is important for teachers to develop a background in alternative approaches to topics. As mentioned earlier, simply rehashing a lesson without any new ideas added may not help a student reach a mastery level. Each teacher is responsible for approaches that are understandable to *every* student, and mastery learning provides an impetus to broaden a teacher's technique.

Testing and diagnosis

Determining how well a student has learned is too broad an issue to detail here. In general, however, evaluating connotes making value judgements. The wisdom of those value judgments improves with the experience of the teacher. The process is improved when the teacher clearly understands what the objectives of instruction are and then evaluates in terms of the objectives. For example, at the community college a ten-item quiz is used to evaluate the students' skill in using percent. Three items involving the skill of stating equivalents in percent, common fraction, and decimal notation reflect an understanding of the concept of percent. Errors on these items usually indicate that the student would benefit from further work with a concrete model of percent. Six items on word problems reflect such skills as being able to set up a proportion from a problem statement, to translate to equation form, to pick out from the problem statement data relevant to the solution, to solve correctly a proportion, to compute accurately, and to use percent that is less than 1 or greater than 100. These skills and the understanding of the concepts behind them are the objectives that must be in the teacher's mind during instruction. As the teacher checks the test and examines what the student has written down, the skills that are lacking become apparent. Remediation then might include help on reading the problem to locate data, computational practice, more concept building on percents less than 1 or greater than 100, or a switch from the equation method to the proportion method of solution. Zeroing in on the particular difficulty means that the student will not have to restudy the whole unit.

The preceding discussion emphasizes paper-and-pencil testing. Fremont (1975) cautions that the student may become "so busy taking tests and trying to do satisfactory work on them that the process of learning seems to become secondary in importance" (p. 324). He advocates diagnosis through the observation of mathematics activities designed to involve a particular skill. Objects attached to a spring (a Slinky toy) spark questions about the amount of stretch, a linear function. The questions can be designed to require whole-number multiplication or division or operations with fractions. He suggests that levers, motion experiments, statistical surveys, probability experiments, and finding the area and volume of familiar objects can form the basis for diagnostic activities. Students who complete the activities have mastered the computational skills required. Not completing the activities leads to self-diagnosis and motivates students by showing them why those skills are useful. An important aspect of the testing and diagnostic procedure arises in the teacher/student interaction. Often mathematically illiterate students behave in a way that increases their chances of failure in a mathematics class. At first, fear of appearing ignorant keeps students from asking questions. As the teacher moves about the room while students are working examples, these students can be identified sitting, pencil in hand, unable to begin the exercise and afraid to ask for help. Helping a student to begin and encouraging a student to ask questions in a one-to-one setting eventually leads to more boldness in asking questions in a group setting. In addition, students who are not learning and have managed to escape the teacher's eye are immediately exposed by the first test. Everyone who scores below the mastery level is invited to select an hour when they can obtain help. (Experience has shown that students are much more responsible about coming in for an appointment with the teacher if they see the teacher writing down their names on an appointment calendar.) The chance to raise a low test score provides incentive for seeking help. The student learns to ask questions, go over missed examples on tests and homework, and, in general, assume an active rather than a passive approach to learning. As the school term progresses, students master one topic after another and meet success on tests. This success and the sense of confidence it inspires seem to give the student more initiative in seeking help and asking questions, behaviors that will be advantageous as the student continues in school.

Sample tests and cumulative tests

In one mastery-learning setting (Haver 1975), students are given for each unit a study guide that includes a sample test for the unit. Students read the book, attend lectures, work problems, and take the sample test before asking for the unit test. Giving students sample tests of a unit focuses their study by defining the behavioral objectives of the unit. Even without the framework of mastery learning, providing sample tests has proved to be effective, since they make clear what is to be learned in the course (Wood 1972).

Including items from previous units on each unit test has also proved to be helpful (Haver 1975). This built-in review system focuses on the cumulative nature of mathematics and contributes to a higher retention of the material as measured by the final examination. The tests are viewed as valuable experience in working problems—each student spends an average of twenty-five hours a quarter working problems. This experience directs the student toward *doing* problems rather than *watching* someone do problems.

Handling papers

In the community college example cited earlier, the teacher implementing mastery learning received no released time for preparation, though extra time would have been helpful for writing and duplicating materials. The teacher prepared each unit just prior to its use. It was helpful to keep notes on needed revisions in the instruction and on the test for subsequent terms. As an example, on the percentage test it was felt that the test items on percentage word problems paralleled the order of the different forms of the problems in the book. Therefore, a student might guess that the first word problem on the test reflected the first form presented in the book, the second word problem represented the second form in the book, and so on. As a result, the test items were rearranged to eliminate this cuing.

Student tests and retests are corrected with the student watching so that the student receives immediate feedback on what needs to be studied. The rest of the class reads introductory material for the next unit while waiting for their tests to be corrected. A file of completed tests is maintained. Since students do not keep the tests, a completely new set of tests and retests is not written every term. All tests from a section of ten to twenty students are filed together. When students wish to use their own test to study for a retest, it is retrieved. Tests are also retained for the entire term as a double-check on any errors made in recording test scores.

An alternative to filing completed tests is to generate ten or more alternative forms of each test. Students may then keep their completed tests for study purposes, since they never know which form they will receive. The amount of work in memorizing the answers to ten alternative tests discourages students from collecting forms of the tests to pass on to other students.

In a setting with several instructors, a testing room can be established so that all instructors have access to study guides, tests, and keys. Another consideration is that grade books must be set up for recording test and retest scores so that it is possible to see at a glance which forms of a unit test a student has already attempted. A good filing and record-keeping system is essential for making effective use of the mastery-learning model.

Students who do not succeed

In general, it appears that implementing mastery learning has led to 70 to 75 percent of the students receiving grades of A and B compared to 40 to 50 percent of the students in a nonmastery format of the same course (Haver 1975; Hector 1975). But whether the grading system assigns C and D grades for effort or F grades for failure to achieve a mastery level, the fact remains that not every student succeeds.

At the community college, to overcome a lack of knowledge about number facts usually requires that the student spend considerable time with flash cards, drill programs on a programmable calculator, or other activities before retaking the course in a subsequent term. More subtle conceptual deficiencies in understanding whole numbers, rational numbers, or place value as described in the case studies by Dahlke (1975) are much harder to overcome. The use of concrete representations such as attribute blocks for classification and set concepts has been helpful in the community college setting. However, one encounters students who attach no meaning to symbols such as "5/8," who persist in following computation algorithms in an inaccurate or rote fashion, or who frequently forget the order in which numbers must be entered into a calculator to use the division key. There is a need for more case studies of individuals who are motivated to learn, are willing to spend time on learning mathematics, and yet are unable to learn with approaches that succeed with the majority of students. Within a given ten-week quarter at the community college, a number of students with only addition and subtraction skills for whole numbers have mastered computation with decimal and common fractions, percentage, and simple algebra. It is true they devoted a considerable amount of time to the task; still, it is not clear what differentiates the student with severe mathematical deficiencies who does achieve a mastery of basic skills from one who does not.

Conclusion

Organizing instruction in a mastery-learning model is one way to reach students who lack basic mathematical skills. A teacher must consider whether the subject matter and skills to be taught are appropriate to the model. The issue of how to vary the time available for achieving mastery is also a major consideration. A teacher employing group-based instruction must consider when time for reteaching is available, how instruction is planned with alternative teaching strategies, what diagnostic testing procedures have been developed, and how papers are handled.

Much of the preceding discussion has focused on how to implement mastery learning. Though the group-based approach is not more costly in financial terms, it does require a commitment of time and effort from the teacher to change from the status quo. Then why would one choose this approach? The answer is simply that more students are given the opportunity to achieve at a higher level. Often a teacher can look at the first test scores in a course and predict who will fail. Neither the teacher nor the students feel good about the failure. Much of that failure could be prevented by organizing instruction to permit those students to be retaught. Increasing the opportunity for reteaching is the main emphasis of mastery learning.

REFERENCES

Allendoerfer, Carl B. "The Second Revolution in Mathematics." *Mathematics Teacher* 58 (December 1965): 690–95.

Bittinger, Marvin L. "A Comparison of Approaches for Teaching Remedial Courses at the College Level." *Mathematics Teacher* 65 (May 1972): 455–58.

Block, James H. "Teachers, Teaching, and Mastery Learning." *Today's Education* 62 (November-December 1973): 30–36.

Block, James H., ed. *Mastery Learning: Theory and Practice.* New York: Holt, Rinehart & Winston, 1971.

Bloom, Benjamin S. "Mastery Learning and Its Implication for Curriculum Development." In *Confronting Curriculum Reform,* edited by Elliot W. Eisner, pp. 17–49. Boston: Little, Brown & Co., 1971.

Brown, O. Robert, Jr. "New Mathematics for a New College." *Mathematics Teacher* 64 (February 1971): 123–26.

Burris, Joanna S., and Lee Schroeder. "Developmental Mathematics: Self-Instruction with Mathematics Laboratory." *Two-Year College Mathematics Journal* 3 (Spring 1972): 16–22.

Cambridge Conference on School Mathematics. *Goals for School Mathematics.* Boston: Houghton Mifflin Co., 1963.

Carroll, John B. "A Model of School Learning." *Teachers College Record* 64 (May 1963): 723–33.

Cole, Blaine L., and Henry S. Weissenfluh. "An Analysis of Teaching Percentages." *Arithmetic Teacher* 21 (March 1974): 226–28.

Dahlke, Richard M. "Studying the Individual in an Individualized Course in Arithmetic at a Community College: A Report on Four Case Studies." *Mathematics Teacher* 68 (March 1975): 181–88.

Dolciani, Mary P., Simon L. Berman, and William Wooton. *Book Two, Modern Algebra and Trigonometry, Structure and Method.* Boston: Houghton Mifflin Co., 1970.

Dolciani, Mary P., and William Wooton. *Book One, Modern Algebra, Structure and Method.* Boston: Houghton Mifflin Co., 1970.

Fremont, Herbert. "Diagnosis: An Active Approach." *Mathematics Teacher* 68 (April 1975): 323–26.

Haver, William E. "Developing Skills in College Algebra—a Mastery Approach." Unpublished manuscript. Knoxville, Tenn.: Mathematics Department, The University of Tennessee, 1975.

Hector, Judith H. "A Mastery Approach to Mathematical Literacy." *Two-Year College Mathematics Journal* 6 (1975): 22–27.

Kipps, Carol. "Who's Committed? Who's Involved?" *Two-Year College Mathematics Journal* 1 (Fall 1970): 32–35.

Kulik, J. A., C. L. Kulik, and K. Carmichael. "The Keller Plan in Science Teaching." *Science* 183 (1974): 379–83.

Mehrens, W. A., and I. J. Lehmann. *Standardized Tests in Education.* 2d ed. New York: Holt, Rinehart & Winston, 1975.

Perry, Donald. "An Experiment in Teaching Elementary Algebra." *Two-Year College Mathematics Journal* 2 (Fall 1971): 40–46.

Riedesel, C. Alan. *Guiding Discovery in Elementary School Mathematics.* 2d ed. New York: Appleton-Century-Crofts, 1973.

Rummell, Frances V. "She Made Mathematics Almost Flunk-Proof." *Education Digest* 24 (October 1958): 32–34.

Spangler, Richard. "Lower Columbia College Mathematics Learning Center." *Mathematics Teacher* 66 (May 1973): 459–62.

Wagner, John, and Howard Jones. "Group-based Instruction: The Best Chance for Success?" *Two-Year College Mathematics Journal* 4 (Winter 1973): 51–54.

Wood, June P. " 'Sample' Tests for Students." *Two-Year College Mathematics Journal* 3 (Spring 1972): 14–15.

Young, D. L., H. E. McKean, and F. L. Newman. "A Personalized System of Instruction in an Undergraduate Mathematics Service Sequence." *American Mathematical Monthly* 81 (August-September 1974): 767–75.

9

Organizing Instruction: Logical Considerations

Thomas J. Cooney

*I*nstruction can be organized in many different ways: arranging for students to work in various-sized groups, providing effective means for discussing homework or doing laboratory activities, or basing a lesson on the logical considerations that arise from the structure of mathematics. This essay will illustrate and elaborate on this latter method.

A theme for organizing instruction will be presented that emphasizes four basic logical aspects of teaching mathematics: (1) recognizing sufficient conditions for specific concepts, (2) recognizing characteristics of specific concepts, (3) selecting examples, nonexamples, and instances, and (4) identifying the conditions that must exist for a mathematical principle to be applied. These four aspects were chosen partly because they have not been treated extensively in the literature and partly because they deserve a higher priority in instruction than they are usually given. Before giving suggestions for organizing instruction based on the four logical considerations above, certain key terms will be defined and illustrated.

Types of Mathematical Knowledge

One of the premises of this essay is that a teacher should realize the existence of different types of mathematical knowledge, such as the following three items:

1. Rectangle

    2. A rectangle is a parallelogram with one right angle.

    3. The diagonals of a rectangle are congruent.

Consider how a teacher might teach these items. For the first item, the teacher could point to specific figures or objects that do or do not represent a rectangle. But this procedure would not make sense with the latter two items. An analysis of the three items shows why. The referent of the first item is a collection of objects having certain attributes; the second and third are truth functional statements, that is, propositions whose truth value can be determined. But the latter two items also differ from each other. The second is the usually accepted definition of a rectangle. As such, it is true by agreement or stipulation. The third statement is true as a result of certain deductions based on previously established or accepted generalizations. Thus a teacher should respond differently to (1) students who do not believe the second item and (2) those who are not convinced that the third statement is true.

A distinction will be made between items of knowledge represented by words or phrases, such as *rectangle,* and items represented by truth functional statements, such as

The diagonals of a rectangle are congruent.

The terms *concept* and *principle* will be used respectively to connote these two types of knowledge. Often, however, these terms are engulfed in semantic confusion in both oral and written contexts. For example, the term *concept* is used in a wide variety of ways, some of which are rather inconsistent. Consider the following illustrations:

- What is your concept of a good proof?
- She used the concept of the distributive law.
- Today we are going to study the concept of the Pythagorean theorem.
- They still haven't learned the concept of $9 + 8 = 17$.

A speaker or author can use a term in any way that he or she deems desirable. But if a term is used in quite different ways, then it must be clarified to prevent misunderstanding. The four illustrations above use the term *concept* in contexts ranging from the somewhat vague construct of good proof to a reference to a basic fact.

Van Engen (1953) and Henderson (1970) have identified some of the problems involved in interpreting the meaning of *concept*. In general, many definitions have been offered by scholars in an effort to clarify the notion of a *concept*. Often the distinctions are subtle and are of little importance for educators. By almost any definition the following terms constitute mathematical concepts: rational number, perpendicular lines, obtuse triangle, prime number, periodic function, polynomial. Each term denotes a concept, and the referent of each term is a set of objects—for example, the set of rational numbers or the set of polynomials. It is this type of concept with which we shall be concerned here.

Objects contained in the referent set of a concept are called *examples,* and objects not so contained are called *nonexamples.* Thus, 2, 3, 5, 7, and 29 are examples of prime numbers, and 4, 6, 9, 25, and 30 are nonexamples.

To avoid confusion, the word *principle* will refer either to a generalization or to a prescription that applies to more than a single mathematical situation. Generalizations state or describe what is the case; prescriptions give directions or advice on how to do something. Examples of generalizations and prescriptions follow.

Generalizations

1. For real numbers x and y, $x + y = y + x$.

2. In a right triangle with sides of length a, b, and c, where c is the hypotenuse, $a^2 + b^2 = c^2$.

3. For real numbers x and y, $(x + y)^2 = x^2 + 2xy + y^2$.

4. If $a|b$ and $b|c$, then $a|c$.

5. Let P denote a polynomial in x. Then $x - a$ is a factor of P if and only if a is a root of the equation $P = 0$.

Prescriptions

1. To solve an equation of the form $ax + b = c$, $a \neq 0$, first subtract b from both members of the equation and then divide each member by a.

2. To round a decimal, add 1 to the digit retained if the first digit dropped is 5 or more. If the first digit is less than 5, leave the retained digit unchanged.

3. To find the area of a trapezoid, multiply one-half the measure of its height by the sum of the measures of the bases.

Just as there are "particulars" for concepts, called *examples,* there are "particulars" for principles which are termed *instances.* An instance of a generalization is obtained by replacing the variables in the generalization with constants from the specified domain. (The domain is often implicit in school mathematics principles.) Thus, the following are instances, respectively, of the generalizations above:

1. $3 + 5 = 5 + 3$.
2. In $\triangle ABC$, $a = 3$, $b = 5$, $c = 4$, $m \angle B = 90$; hence, $3^2 + 4^2 = 5^2$.
3. $(\pi + {}^-7)^2 = \pi^2 + 2(\pi)({}^-7) + ({}^-7)^2$.
4. $2|6$ and $6|12$; hence, $2|12$.
5. 3 is a root of $x^3 - 2x^2 - 4x + 3 = 0$ if and only if $x - 3$ is a factor of $x^3 - 2x^2 - 4x + 3$.

An analogous situation exists with respect to prescriptions. Each of the three prescriptions above implicitly involves a domain over which the prescription applies. If constants are selected from the respective domains and the indicated procedure is carried out, then the result will be called an instance of the prescription.

It should be noted that whereas principles are truth functional, concepts are not. This difference alone suggests that the ways in which these two types of mathematical knowledge are taught should be different. Having stipulated how key terms will be used, we can now discuss the four basic considerations stated earlier.

Recognizing Sufficient Conditions for Specific Concepts

A fundamental problem that students have in learning mathematics is failing to recognize the conditions sufficient for an object to exemplify a concept. Sometimes, the conditions sufficient for an object to be in a concept's referent set may be identified simply by recalling the definition of the term denoting the concept and pointing out the biconditional nature of the definition. For example, if students are asked to prove that a figure is a rhombus and they cannot recall how to prove it, then recalling the definition of *rhombus* and observing necessary and sufficient aspects of the definition might suffice. Students are frequently unaware that a definition involves conditions necessary and sufficient for an object to be an example of a concept. The following definition

A *rhombus* is a parallelogram all of whose sides are congruent

gives rise to two conditional statements:

1. If a figure is a parallelogram all of whose sides are congruent, then it is a rhombus (sufficient condition).
2. If a figure is a rhombus, then it is a parallelogram all of whose sides are congruent (necessary condition).

Therefore, *when teaching a concept, make students explicitly aware of the sufficient condition implied by the definition of the concept.*

However, sufficient conditions for concepts are not limited to those contained in definitions. For example, students are sometimes unsure about whether triangles are similar or about what conditions must be met to have a periodic function, an incenter of a triangle, or equivalent equations. The following dialogue typifies learning difficulties of this sort:

> *Teacher:* Now we have to prove that in similar triangles corresponding medians have the same ratio as corresponding sides. In terms of our diagram here [*fig. 9.1*], what does that tell us, Bob?
>
> *Bob:* Well, triangle *ABC* is similar to triangle *DEF,* and segments *BG* and *EH* are medians.
>
> *Teacher:* And what do we have to show?
>
> *Bob:* That the ratio of the lengths *BG* and *EH* is the same as the ratio

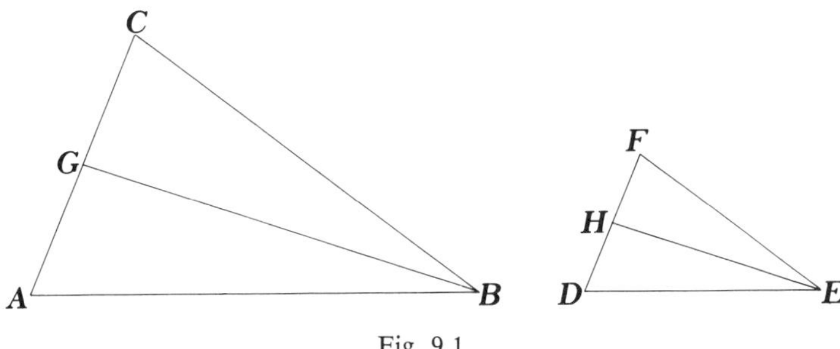

Fig. 9.1

of the lengths *BC* and *EF*—or any other pair of corresponding sides.

Teacher: All right, if we could show that the two ratios are equal, we should have it. How can we do that? Lori?

Lori: Usually we do things like that with similar triangles.

Teacher: Sounds like a good approach. Which triangles should we consider?

Geneva: Probably triangles *BCG* and *EFH*.

Teacher: And how can we prove those two triangles similar?

Alex: I suppose by getting angles congruent.

Teacher: Can you be a little more precise?

Alex: Well, if two angles of one triangle are congruent to two angles of another, then the triangles are similar.

Teacher: Do you think that method would work here? [*After some discussion, the students conclude that it would not.*] Do we have any other methods?

Curtis: How about sides congruent—I mean proportional. Didn't we have a theorem about that?

Teacher: Yes, we did. Would that help us? [*Students think it might but are not sure.*] Well, do we have any other means of proving triangles similar?

Students: I don't think so.

Apparently the sufficient condition stemming from the definition (at least the usual definition given in high school geometry) was not the source of the problem. Rather, the students seemed unaware of other conditions also sufficient for concluding that triangles are similar—in particular, the condition usually referred to as the SAS similarity theorem. Sometimes a brief reminder may be enough to resolve such a problem, but sometimes the student does not realize that a theorem can be used to establish that an object is an example of a specific concept. Teachers should emphasize the variety of ways a referent set can be characterized.

To illustrate the problem further, consider a sample of student work that typifies the errors sometimes made in solving equations:

$$x^3 - x^2 - 2x = 0$$
$$x^2 - x - 2 = 0$$
$$(x - 2)(x + 1)$$
$$x = 2, -1$$

Although several misunderstandings might have contributed to the deletion of 0 as a root, one plausible explanation is that the student did not know the conditions or operations that produce equivalent equations. The learning problem might be resolved by asking the student what means are available for producing equivalent equations (assuming he or she knows what equivalent equations are) or by asking the student to consider replacing x by 0 in the original equation and to observe the resulting statement. Once the student has discovered that the manipulation of the original equation was not justified, the question might be posed, "What techniques do we have available for insuring that an equivalent equation is produced when we solve equations?" The student should be assisted in organizing the various techniques for generating equivalent equations. Hence, the following suggestion is offered to assist in planning instruction: *Help students categorize all the theorems or properties that state a sufficient condition for an object to be in the referent set of a concept.*

Sufficient conditions for concepts are also embedded in set inclusion relationships. Consider the diagrams involving the mathematical concepts of *complex number* and *parallelogram* depicted in figures 9.2 and 9.3. It follows that being an integer is sufficient for being a rational number, which is sufficient for being a real number, which in turn is sufficient for being a

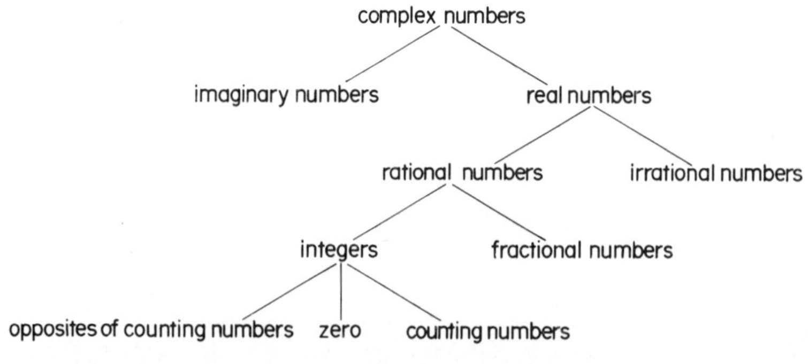

Fig. 9.2

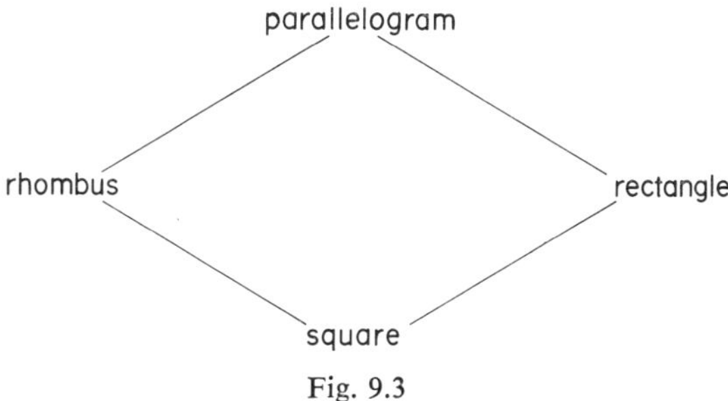

Fig. 9.3

complex number. Likewise, being a square is sufficient for being a rectangle, which in turn is sufficient for being a parallelogram.

It is important for students to understand set inclusion relationships in order to understand the structure of mathematics and to be able to categorize as much information as possible. The advantage of knowing set inclusion relationships is in the inferences they allow one to draw. That is, any property that holds for every element in the more inclusive set must also hold for every element in any subset of the larger set. Thus, since every integer is a rational number, any property that holds for every rational number must also hold for every integer. (It should be noted, however, that a property that holds for the rational number system, as opposed to a property that holds for every rational number, does not necessarily hold for the system of integers. To illustrate, consider the property of density.) The drawing of inferences is explored more extensively in the next section. Let it suffice here to say that knowing certain set inclusion relationships facilitates the organization of one's information. Hence, *where set inclusion relationships exist with respect to concluding that an object is an example of a concept, organize instruction to identify and illustrate those relationships.*

The conclusion of this section is that one aspect of organizing instruction lies in focusing on the sufficient conditions that allow one to conclude that an object exemplifies a concept. These conditions may be contained in a definition, a set inclusion relationship, or some other outcome of established mathematical principles.

Recognizing Characteristics of Specific Concepts

Another basic conceptual problem for students lies in not being able to identify all the information possible when given an example of a concept. The following dialogue illustrates this problem:

Teacher: Now let's use this figure [*fig. 9.4*] and see if we can find a formula that will give us the area of rhombus *ABCD* in terms of the

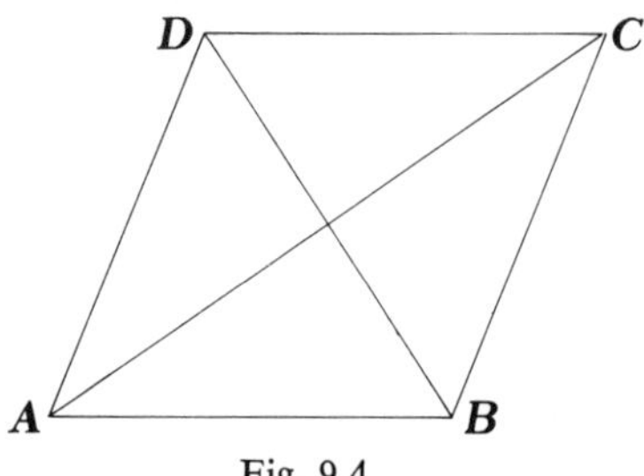

Fig. 9.4

diagonals *AC* and *DB*. How can we start? [*No response.*] What do we know about the diagonals of a rhombus, Hank?

Hank: Well, I think we had a theorem that says they are perpendicular.

Teacher: Yes, we did. Can you tell us anything else?

Hank: I don't think we had any other theorem about it.

Judy: Won't they bisect each other also?

Fred: I don't remember a theorem about that.

Teacher: Well, Fred, it is true. But how do we know that?

Judy: Well, it's true for a parallelogram.

Hank: But this is a rhombus—we don't have any theorem about the diagonals of a rhombus bisecting each other.

Hank's difficulty relates to the problem of making inferences about concepts related by set inclusion relationships. He was probably right when he declared that the class had not studied a theorem stating that the diagonals of a rhombus bisect each other. Hank's problem might have been avoided had previous lessons been organized to maximize the identification of such knowledge.

A similar situation arises when one considers the relationship between the set of continuous functions and the set of differentiable functions. The latter is a subset of the former. It follows that any property true for all continuous functions must also be true for differentiable functions. Hence, the theorem that establishes that in a closed interval a continuous function has a greatest and a least value allows one to conclude that this same property holds for differentiable functions. Likewise, other properties of continuous functions are also applicable to differentiable functions, although such properties are seldom explicitly stated in textbooks. Textbook writers do not state these properties because they assume readers are capable of drawing the needed inferences. Such assumptions are reasonable, but teachers need to be alert to evidence that students are not recognizing the relationships.

Teachers of trigonometry can attest that students do not realize how a property of periodic functions plays a role in solving equations such as this one:

$$\sin x = 0$$

Some students will maintain that the solution set is $\{0, \pi\}$. They fail to realize that the sine function is periodic and that therefore the solution set is necessarily infinite, not finite. This problem, too, can be seen as an instance of students failing to understand the set inclusion relationship and the inferences that it allows one to draw.

It is important to have students understand set inclusion relationships and how these relationships can facilitate the organization of one's knowledge. To assist students in organizing their knowledge, *help them identify properties of objects that exist because of set inclusion relationships.*

Not all the knowledge one can infer from a concept, however, is embedded in set inclusion relationships. Some knowledge exists because of other types of relationships. Consider the following dialogue:

Teacher: Now let's see if we can identify which numbers in base five are the even numbers. Does anyone think they might know?

Andy: It would be 0, 2, 4, 10, 12, 14, 20, and so on.

Teacher: And why do you say that?

Andy: Because even numbers always end in 0, 2, 4, 6, or 8 but in base five we don't have a 6 or an 8.

Teacher: Yes, it is true that the even numbers in the decimal system end in 0, 2, 4, 6, or 8, but could someone give me a different idea of what an even number is? [*Long pause. Finally one girl raises her hand timidly.*] Anna?

Anna: Well, they are multiples of two—but that's the same thing—2, 4, 6, 8, and so on.

Teacher: Let's look at that. Multiples of two. What does that mean? Frank?

Frank: $2, 2 \times 2, 3 \times 2, 4 \times 2$, and so on.

Teacher: In terms of addition how would you express that?

Frank: $2, 2 + 2, 2 + 2 + 2, 2 + 2 + 2 + 2$.

Teacher: Exactly. So what has Frank represented? Kelly?

Kelly: Even numbers—but I'm not sure how that helps.

Kim: Oh, I think I see. The even numbers are $0, 2, 2 + 2, 2 + 2 + 2, 2 + 2 + 2 + 2$, and so on, but in base five those numbers aren't written as 0, 2, 4, 6, 8, . . .

Teacher: Very good. How are they written?

Kathy: Let's see, 0, 2, 4 [*thinks a little longer*], 11, 13, 20, and so on.

In this situation it was not that the students had forgotten knowledge about a concept but rather that their existing knowledge was not broad enough. In particular, their notion of an even number was locked into a particular representation, namely, the decimal system. The teacher's strategy was to broaden their understanding by getting them to focus on a property of the even numbers, not the representation. That is, the teacher wanted them to have more information—a better understanding, if you will—about the concept of even number.

To illustrate further, consider Rob, a junior high school student who, when asked to find the square root of 55 225, obtains the correct result, 235. When asked to find the product of 235 and 235, however, he says he is not sure and resorts to carrying out the computation. Why does he resort to multiplying in order to answer the question? A reasonable explanation is that his understanding of square root is not very thorough. To put it another way, he is not able to derive relevant information about the given concept. To emphasize the importance of organizing instruction to promote such knowledge, the following suggestion is given. *Make a point to identify and review basic properties of objects that exemplify key mathematical concepts.*

To summarize the preceding discussion, it is important that students be able to make justifiable inferences about concepts. Further, instruction should be organized to encourage such inferences by emphasizing set inclusion relationships and by providing students with activities aimed at developing a thorough conceptual understanding. This section and the preceding one can provide a coordinated basis for promoting conceptual development and organizing knowledge concerning concepts. The two sections relate to the teaching and learning of sufficient and necessary conditions for an object to be an example of a mathematical concept. Although the mode of instruction may range from the lecture method to the laboratory method, two questions can serve as a central focus for planning instruction:

1. What conditions must we have before we can be sure we have a □? (When □ stands for a specific concept.)
2. Now we have a □. What does that tell us?

Selecting Examples, Nonexamples, and Instances

Another consideration in organizing instruction is the selection of particular examples, nonexamples, and instances. In selecting "particulars," the teacher has several factors to consider:

1. How should examples, nonexamples, or instances be sequenced to maximize understanding?
2. Should examples, nonexamples, or instances be presented first, followed by the "discovery" of the concept or principle? Or should the concept or principle be identified first, followed by the presentation of examples, nonexamples, or instances?

3. Should examples, nonexamples, or instances be selected by the teacher or by the students?

Each of these questions will be discussed. But first the role that examples and nonexamples play in how students learn concepts will be considered.

For an object to be contained in the referent set of a concept, it must satisfy specified conditions related to the defining attributes of the concept. It follows that any object that satisfies all the conditions is an example of that concept. A nonexample is some object that satisfies fewer than all the conditions. To illustrate, consider the concept of function. If an object satisfies the following two conditions, it is a function:

1. It is a relation.
2. Each element in the domain of the relation has a unique image in the range.

Examples must satisfy both conditions, but nonexamples can satisfy at most one of the conditions. Thus the relation defined by the equation

$$x^2 + y^2 = 10, \text{ where } x \text{ and } y \text{ are real numbers,}$$

is a nonexample of a function, since it does not satisfy the second condition. A chair and the expression $x + 2y$ are also nonexamples of a function because they fail to satisfy either of the two conditions. In selecting nonexamples, a teacher should decide which conditions will not be met. Generating nonexamples by omitting one condition at a time can help insure that a wide variety of nonexamples are selected.

Each example of a concept must necessarily possess the defining attributes of that concept. But each example also contains irrelevant attributes. Menchinskaya (1969) emphasized that examples should be selected in such a way that the irrelevant attributes are varied so they are not mistakenly thought to be relevant attributes. Zykova (1969) noted that the visual representation of examples is a factor in influencing what the student perceives to be relevant attributes. She observed that some sixth graders thought a circle had exactly two diameters, one vertical and one horizontal. Similarly, some pupils considered the exterior angle of a triangle to be an obtuse angle placed to the right of its supplementary angle. However, neither an obtuse angle on the left nor an acute angle on the right were considered to be examples of an exterior angle of a triangle. In short, the students failed to note several attributes that are irrelevant for an object to be an exterior angle of a triangle. This misconception probably had its basis in the limited examples presented to the students.

In teaching functions, a teacher might have students examine graphs of functions and nonfunctions as shown in figure 9.5. By comparing and contrasting the graphs, the students could identify a defining attribute of functions, namely, that each element in the domain must have a unique image.

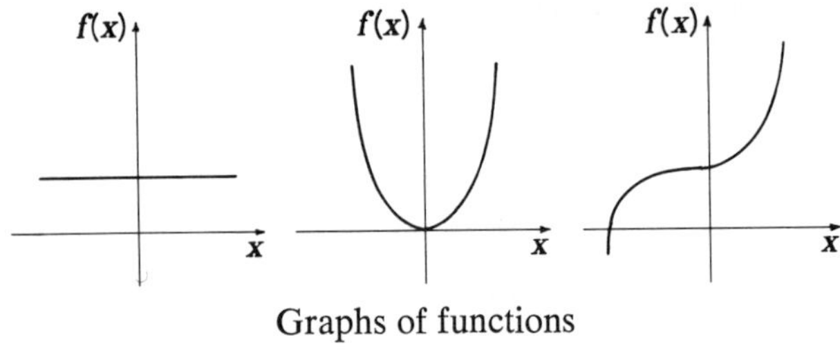

Graphs of functions

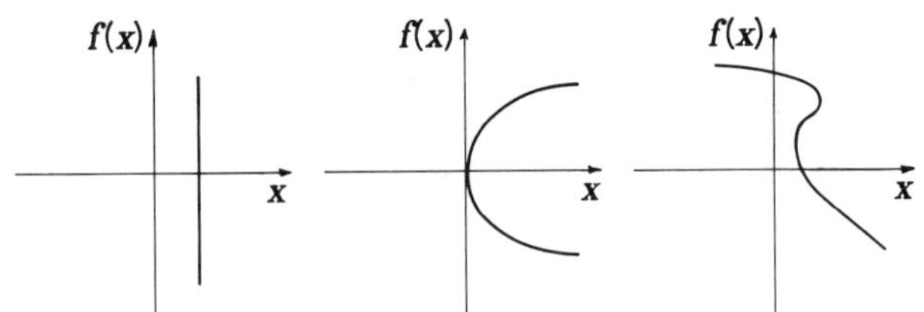

Graphs of relations that are not functions

Fig. 9.5

But these examples do not allow students to determine if the graph of a relation must be a continuous curve in order to be a function. Further, can a finite relation constitute a function? This and other questions should be addressed in selecting examples and nonexamples.

A definitive statement concerning the relative effectiveness of various sequences of nonexamples alone or in combination with examples is not justified. The teacher should plan, however, to select a sequence that enables the learner to distinguish between examples and nonexamples; this entails being able to recognize objects possessing the defining attributes of a concept as opposed to those that do not.

Of course, not all the examples and nonexamples need to be presented or created by the teacher. Students can also generate examples and non-examples for consideration. A student's ability to identify examples and nonexamples of a concept is one measure of an understanding of that concept. For example, a teacher might pose the following questions when teaching the concept of reciprocal:

1. Can you give me an example of a number whose reciprocal is larger than the given number? Equal to the given number? Less than the given number?

2. Can you identify a number whose reciprocal does not exist?

To illustrate further how defining attributes can be teased from irrelevant factors, consider the concept of trapezoid and the following instructions from the teacher:

1. Give an example of a quadrilateral with exactly one line of symmetry, that line not being a diagonal.

2. Give an example of a quadrilateral with a pair of parallel sides but with opposite angles that are not congruent.

3. Give an example of a figure that is not a trapezoid but that has exactly one pair of parallel sides.

One consideration in sequencing examples and nonexamples is to help students realize the extensiveness of a concept's referent set. Suppose a teacher, discussing irrational numbers, presents only the examples $\sqrt{2}$, $\sqrt{3}$, $\sqrt{7}$, $\sqrt{11}$, and $\sqrt{19}$. Further, suppose that when the students are asked to give examples of irrational numbers, they give $\sqrt{5}$, $\sqrt{7}$, $\sqrt{13}$, and $\sqrt{17}$. What conclusion can be drawn with respect to their understanding of irrational number? Clearly they can generate irrational numbers. But do they believe that an irrational number must be of the form $\sqrt{p}$, where p is a prime? How might they classify numbers such as $\sqrt{6}$, $\sqrt{12}$, $\sqrt{26}$, $\sqrt{18}$, and $\sqrt{21}$? How would they classify numbers such as $\sqrt{4}$, $\sqrt{9}$, and $\sqrt{16}$? Or $\sqrt[3]{3}$, $\sqrt[4]{7}$, and $\sqrt[5]{16}$. Would they think that any number is irrational if it can be expressed in the form $\sqrt[n]{N}$, where $N>0$ and n is a positive integer greater than one? Of course, the answers to these questions cannot be determined without evidence from students' written work or oral responses. The point is, however, that by presenting a narrow selection of examples and nonexamples, a teacher runs the risk of giving students a restricted notion about the particular concept (and its referent set). Hence, *select and organize examples and nonexamples so that students can more easily identify relevant and irrelevant attributes of the concept to be learned and can realize the extensiveness of a concept's referent set.*

Consider now the organizing and sequencing of instances. Instances can play at least two different roles in the teaching of principles. First, they can help clarify the meaning of a principle for students; second, they can provide a basis for believing a principle is true. These two purposes are described and illustrated by Cooney, Davis, and Henderson (1975). The following generalization can be used to illustrate the purposes:

The square of any odd number less one is divisible by 8.

An instance of this generalization can be shown by selecting an odd number, say, 7, and observing that

$$7^2 - 1 = 48 \text{ and that } 8 \mid 48.$$

One student might react to this instance by saying, "Oh, I see what it means now." Another might say, "By golly, it does work." For the first student the instance apparently clarified the generalization, but for the second it appeared to provide evidence that the generalization is true. Sometimes a teacher cannot tell whether an instance of a generalization has served one purpose or the other. The purpose may depend on the student involved.

In selecting instances, teachers should consider these two purposes. If instances are to serve the purpose of clarification, then the selection should involve constants that are familiar to students. For the generalization above, the number 7, along with other smaller positive odd numbers, would probably suffice. But to convince students a generalization is true, instances should be selected that represent various subsets of the domain of the generalization. For example, instances with numbers such as $-51, -1, +1$, 63, and 127 should be used. The selection of instances for clarifying or justifying a generalization should be given careful consideration when a lesson is organized. Such an important teaching decision should not be made casually. Further, the use of instances is one of the few means of justifying mathematical principles to students who are relatively immature mathematically.

Many of the same remarks apply to the teaching of prescriptive principles. An instance of a prescription can be thought of as a demonstration of the procedure indicated by the prescription. Consider this prescription:

> To form the inverse of a nonsingular matrix A, replace each element a_{ij} of A by its cofactor, divide each element of the resulting matrix of cofactors by the determinant of A, and transpose the result.

An instance of the prescription is as follows:

Let

$$A = \begin{bmatrix} 1 & 0 & 2 \\ 6 & -1 & 3 \\ -2 & -4 & 5 \end{bmatrix}.$$

The matrix of cofactors is

$$\begin{bmatrix} 7 & -36 & -26 \\ -8 & 9 & 4 \\ 2 & 9 & -1 \end{bmatrix}.$$

Since $|A| = -45$, then

$$A^{-1} = \begin{bmatrix} \dfrac{7}{-45} & \dfrac{-36}{-45} & \dfrac{-26}{-45} \\ \dfrac{-8}{-45} & \dfrac{9}{-45} & \dfrac{4}{-45} \\ \dfrac{2}{-45} & \dfrac{9}{-45} & \dfrac{-1}{-45} \end{bmatrix}^T$$

$$A^{-1} = \begin{bmatrix} \dfrac{7}{-45} & \dfrac{-8}{-45} & \dfrac{2}{-45} \\[2ex] \dfrac{-36}{-45} & \dfrac{9}{-45} & \dfrac{9}{-45} \\[2ex] \dfrac{-26}{-45} & \dfrac{4}{-45} & \dfrac{-1}{-45} \end{bmatrix}$$

or

$$\begin{bmatrix} -\dfrac{7}{45} & \dfrac{8}{45} & -\dfrac{2}{45} \\[2ex] -\dfrac{4}{5} & -\dfrac{1}{5} & -\dfrac{1}{5} \\[2ex] \dfrac{26}{45} & -\dfrac{4}{45} & \dfrac{1}{45} \end{bmatrix}.$$

Demonstrations of a procedure enable students to follow the procedure on their own. In this sense, instances are used to help clarify the procedure. If a teacher wants to establish the correctness of a prescription, additional demonstrations will not suffice. One must either resort to a deductive argument or show that following the prescription leads to a desired result, for example, showing that the product of A and A^{-1} is the identity matrix for multiplication. Thus instances can both clarify and justify a generalization, but they serve only to clarify a prescription.

Like examples and nonexamples, instances can be selected and proposed by students as well as by teachers. Surely an argument aimed at justifying a principle is more credible to students if they select the instances. Further, in clarifying a procedure, teachers may want students to do several demonstrations, either at their seats or at the chalkboard.

Instances can form the basis of a discovery lesson if they are presented so that the students must identify the principle by inductive reasoning. They can also serve as the primary focus to be considered in deciding whether or not to plan a discovery lesson. The generalization earlier concerning divisibility by eight, the Pythagorean theorem, and $(x + y)(x - y) = x^2 - y^2$, where x and y are real numbers, are all principles that are amenable to teaching by discovery. Principles that are complex, however—for example, the half-angle trigonometric identities—are probably best taught by exposition with the principle presented first and then instances for clarifying and justifying it.

To conclude this section, the following suggestion is given: *In planning for the identification of instances (by either yourself or the students), be sure that the selection is representative of the domain to which the generalization applies.*

Analyzing Mathematical Principles

One of the most important aspects when teaching principles is to emphasize when a principle can be used and, if it is to be proved, what can be assumed and what must be proved. Failing to understand the conditions under which a principle can be used is a common problem for mathematics students. Consider the following situation, in which the quadratic formula is being taught.

Teacher: Ellen, what did you get when you solved $x^2 - 10x = 5$?

Ellen: $x = \dfrac{10 \pm 4\sqrt{5}}{2}$, or $5 \pm 2\sqrt{5}$.

Byron: That's not what I got.

Teacher: What did you get?

Byron: $x = 5 \pm \sqrt{30}.$

Teacher: That's right. Ellen, what values for a, b, and c did you use in the quadratic formula?

Ellen: $a = 1$, $b = -10$, and $c = 5$.

Ellen's mistake probably occurred because she neglected one of the conditions of the quadratic formula, namely, that the equation be in the form

$$ax^2 + bx + c = 0.$$

A similar situation arises when students maintain that, in figure 9.6, $m < 1 = m < 2$ because $AB = AC$. Their problem is similar to Ellen's in that they have applied a principle without making sure they have met the conditions necessary for using it. The students may forget or may never have been aware of the condition that the congruent sides and angles must be in the same triangle. This condition is usually an implicit part of what is often referred to as the *isosceles triangle theorem.*

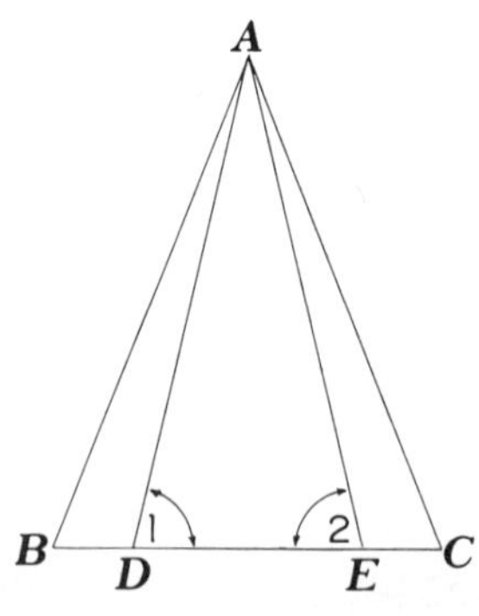

Fig. 9.6

There are several ways a teacher can help students identify situations in which a principle applies. To illustrate, consider the Pythagorean theorem. A teacher might pose the question, "In which of the following four situations could the Pythagorean theorem be used to solve the problem?"

1.
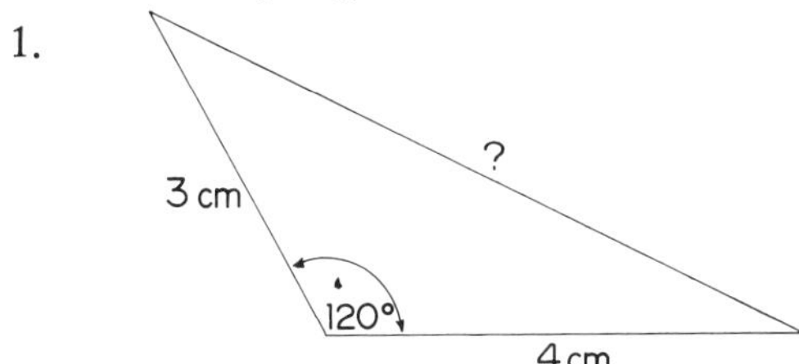

2. What is the longest diagonal possible in a rectangular prism of dimensions 4 m $\times$ 8 m $\times$ 9 m?

3. Find the area of a right triangle if the hypotenuse is 13 cm long and one leg is 5 cm long.

4. If the dimensions of a triangle are 6 m, 8 m, and 10 m, is the triangle a right triangle?

The answer to the teacher's question is situations 2 and 3. In both these problems the Pythagorean theorem can be used to find the answer. In situations 1 and 4, the Pythagorean theorem cannot be used, but the reason is different in each situation. In 1, the triangle is not a right triangle and hence fails to satisfy the condition for using the theorem. In 4, the question is asking for an inference based on the converse of the Pythagorean theorem and not on the theorem itself.

Another teaching technique is to present students with a situation or problem and ask them what principle is needed to solve the problem. For example, one could present a student with a situation that calls for finding the volume of a right circular cone and ask which mathematical principle could be used to find the volume. Or a student could be asked to determine the sum of the two roots of a quadratic equation without first finding the individual roots. In situations such as these the student is asked, not to produce a numerical answer, but to indicate which principle could be used.

Still another technique would be to present students with a situation and ask them to determine if a certain principle has been applied appropriately. To illustrate, suppose students are asked to consider the following principle:

$$\text{For } a > 0, b > 0, \sqrt{a} \cdot \sqrt{b} = \sqrt{ab}$$

Then they are asked to consider this statement:

$$\sqrt{-1} \cdot \sqrt{-4} = \sqrt{(-1) \cdot (-4)} = \sqrt{+4} = 2$$

The following questions might be posed:

- Is the statement true?

- Can the statement be justified on the basis of the principle above?
- Why or why not?

To answer these questions, the student must first determine the conditions required for the principle to be used. In this example, a and b must be positive, which they are not in the given statement. *In general, when teaching a principle, have students identify situations in which the principle does apply and situations in which it does not apply.*

Using the Suggestions in a Variety of Instructional Contexts

The foregoing suggestions apply to organizing instruction for any instructional mode, for example, expository, discovery, or laboratory. Several specific ways for using the suggestions in a variety of classroom situations are now offered.

Most middle school students enjoy playing the game Concentration. Consider the following adaptation, which entails examples of concepts. A four-by-four array of squares is usually appropriate. The squares could contain names of concepts and examples, as illustrated in figure 9.7 for use in a

Fig. 9.7

unit on number theory. The goal is to uncover matching squares, for example, the squares "2" and "even prime." The same game could be used for teaching geometry by using pictures of figures matched with either the names of the figures or the definitions of the related concepts.

As mentioned in a preceding section, instances can be used in teaching by discovery. For example, suppose you are teaching what is sometimes called the *rational root theorem*. The following three equations could be presented to students for them to solve:

1. $x^3 - 6x^2 + 11x - 6 = 0$
2. $x^3 - 9x^2 + 23x - 15 = 0$
3. $x^3 - 7x^2 + 14x - 8 = 0$

Assume that they solve the equations correctly and obtain the following roots, respectively.

1. $r_1 = 1, r_2 = 2, r_3 = 3$
2. $r_1 = 1, r_2 = 3, r_3 = 5$
3. $r_1 = 1, r_2 = 2, r_3 = 4$

Students might conjecture that the roots of polynomial equations with leading coefficient 1 are integral factors of the constant term or that all cubic equations have 1 as a root. If students consider these conjectures, an equation such as the following could be presented:

$$2x^3 + 3x^2 - 32x + 15 = 0$$

Similar instances might lead students to conclude that the roots are related to the factors of the leading coefficient and the constant term. The question might then arise whether factors of these terms might be used to state a necessary or a sufficient condition for a number to be a root of an integral polynomial equation. For students to define the domain of a generalization such as the rational root theorem, it is important that instances be selected that provide information about that domain.

The suggestion on selecting examples and nonexamples can also serve as a basis for a discovery lesson. For example, after investigating selected examples, students might discover that perfect squares end (that is, the unit's digit) only in 0, 1, 4, 5, 6, or 9. These questions might then be posed:

1. Does every perfect square end in 0, 1, 4, 5, 6, or 9?

2. If a number ends in 0, 1, 4, 5, 6, or 9, must it be a perfect square?

To emphasize that the property under consideration is a necessary but not a sufficient condition for a natural number to be a perfect square, the following nonexamples might be presented: 50, 21, 44, 15, 26, and 39.

The four considerations discussed previously can also serve as an organizing theme for creating questions to pose for students. The questions could be used for purposes of motivation, diagnosis, or evaluation. Possible questions reflecting the different logical considerations and the related suggestions are illustrated below. Assume the questions are intended for an average or above-average class of seventh graders studying a unit on equivalent fractions. Of course, the questions might be posed better orally.

Recognizing sufficient conditions for specific concepts

1. Identify two ways in which you can tell whether two fractions are equivalent.

2. Mark the following statements yes or no, depending on whether the two indicated fractions are equivalent:

   a) The numerator and denominator of the first fraction are each multiplied by 2 to get the second fraction. (Yes or No)

   b) The number 2 is added to both the numerator and denominator of the first fraction to get the second fraction. (Yes or No)

   c) If the first fraction is multiplied by 2, you get the second fraction. (Yes or No)

Recognizing characteristics of specific concepts

1. The two circles below are the same size. The shaded portions represent the same fraction of each circular region. Does circle *A* or circle *B* have the larger region shaded, or are they the same?

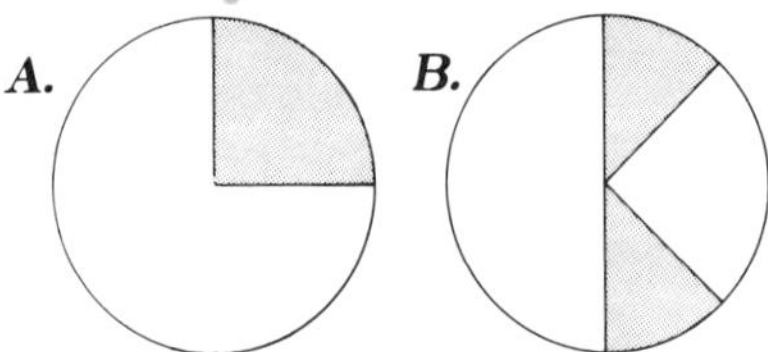

2. Two fractions are equivalent. The first fraction is represented by point *a* on the number line below. The denominator of the second

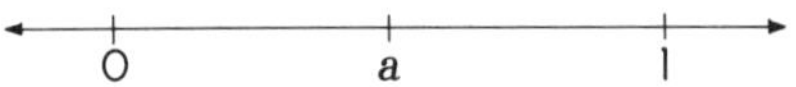

   fraction is twice as large as the denominator of the first fraction. Where would the second fraction be represented on the number line?

Selecting examples, nonexamples, and instances

1. Is it possible for two fractions to be equivalent and have different denominators? If so, write two such fractions.

2. Is it possible for two fractions to have the same numerator and not be equivalent? If so, write two such fractions.

3. The shaded portion of the rectangle below represents ⅓ of the entire region of the rectangle.

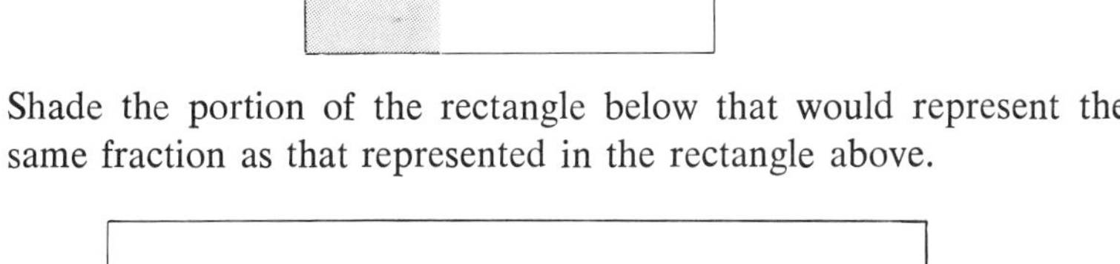

Shade the portion of the rectangle below that would represent the same fraction as that represented in the rectangle above.

These questions focus on ascertaining the student's acquisition of the concept of equivalent fractions. The fourth consideration, analyzing mathematical principles, can also serve as a basis for generating questions. This was illustrated earlier using the Pythagorean theorem. Each of the four considerations and the related suggestions can serve as a means of identifying key questions around which instruction can be organized.

The same four ideas can also provide a means by which review lessons can be organized. In reviewing a particular concept, teachers should make sure that all the sufficient conditions for identifying examples of that concept are identified. Further, all the properties possessed by examples of the concept ought to be made explicit. It should be made clear what constitutes examples and nonexamples of a given concept. Instances of a given principle and situations when the principle can be applied should also be identified. Some teachers like students to keep notebooks for purposes of study and review. The suggestions given in the preceding sections could serve as an organizing theme for the notebooks.

Conclusion

The object of any organizational theme in teaching mathematics is to facilitate the students' learning of mathematics. Students' ability to learn mathematics is directly related to their understanding of how mathematical concepts and principles are related—in short, an understanding of the structure of mathematics. The four logical considerations that have been discussed here are intended to highlight basic relationships that exist in mathematics. The major thesis of this essay is that these considerations ought to receive primary attention in any strategy a teacher uses for organizing classroom instruction.

REFERENCES

Cooney, Thomas J., Edward J. Davis, and Kenneth B. Henderson. *Dynamics of Teaching Secondary School Mathematics.* Boston: Houghton Mifflin Co., 1975.

Henderson, Kenneth B. "Concepts." In *The Teaching of Secondary School Mathematics,* edited by Myron F. Rosskopf, pp. 166–95. Thirty-third Yearbook of the National Council of Teachers of Mathematics. Washington, D.C.: The Council, 1970.

Menchinskaya, N. A. "The Psychology of Mastering Concepts: Fundamental Problems and Methods of Research." In *The Learning of Mathematical Concepts,* edited by Jeremy Kilpatrick and Izaak Wirszup, pp. 75–92. Soviet Studies in the Psychology of Learning and Teaching Mathematics, vol. 1. Stanford, Calif.: School Mathematics Study Group, 1969.

Van Engen, Henry. "The Formation of Concepts." In *The Learning of Mathematics: Its Theory and Practice,* edited by Howard F. Fehr, pp. 69–98. Twenty-first Yearbook of the National Council of Teachers of Mathematics. Washington, D.C.: The Council, 1953.

Zykova, V. I. "The Psychology of Sixth-Grade Pupils' Mastery of Geometric Concepts." In *The Learning of Mathematical Concepts,* edited by Jeremy Kilpatrick and Izaak Wirszup, pp. 149–88. Soviet Studies in the Psychology of Learning and Teaching Mathematics, vol 1. Stanford, Calif.: School Mathematics Study Group, 1969.

10

Organizing for Alternative Schools: An Integrated Curriculum

Jane Donnelly Gawronski

Joan E. Fehlen

All that has been said of the importance of individuality of character, and diversity in opinions and modes of conduct, involves, as of the same unspeakable importance, diversity of education. A general state education is a mere contrivance for molding people to be exactly like one another. And as the mold in which it casts them is that which pleases the predominant power in the government, whether this be a monarch, a priesthood, an aristocracy, or the majority of the existing generation in proportion as it is efficient and successful, it establishes a despotism over the mind, leading by national tendency to one over the body. An education established and controlled by the state should only exist, if it exist at all, as one among many competing experiments, carried on for the purpose of example and stimulus, to keep the others up to a certain standard of excellence.

John Stuart Mill: *On Liberty* (1859)

*T*he growth of alternatives and options in public education is an acknowledgment that no *one* method of education is best for all children. Children are different and can retain their individuality best in a learning environment that is tailored to their differences rather than one that requires all individuals to conform to one educational model. Children learn at different rates and in different ways. In fact, the same child may learn in different ways at different stages of development. A variety of models of education, then, is necessary to meet the needs of children having a variety of learning styles.

Having alternatives in education means having a choice among different educational models. In an alternative school setting, pupils, parents, and teachers each choose what they feel is best. There is no intervention, no coercion—only voluntary choice (North Central Association of Colleges and Schools 1974–75, p. 1).

Before alternatives or options in education can be considered available, the following must be true:

1. More than one educational model must be available; that is, there must be a choice.

2. Parents, pupils, and teachers must be free to choose an alternative. One model should not be superimposed or mandated over another for any reason.

3. All available alternatives or schools must be considered equally legitimate by the school system and the clientele it serves.

4. Equal access to alternatives must be guaranteed to all. No one should be excluded by reason of achievement, special needs, geography, or other factors. Access should not be based on a status system of human classification.

5. The school (staff and pupils) must show evidence of working toward the attainment of a comprehensive set of educational objectives.

6. The types of educational models and their goals must be communicated accurately and freely to everyone involved—parents, pupils, teachers.

7. The program must operate on a full-time basis, not just for part of a day.

One alternative that meets these criteria is the open school. The open school has been defined in several ways. One community that developed the open-school option defines the philosophy of its school as follows:

> Open education means providing learning experiences that allow an individual to develop his (her) own particular talents and strengths. School activities should be an outgrowth of the interests of children, using the entire community as a laboratory for learning.
>
> Children should learn to read, write and solve problems without going through a structured and sequential skill-building curriculum. Space arrangement, group formation, and scheduling should remain flexible.

Learning should be as individualized and personalized as possible, but each child also should engage in a variety of flexible groupings in order to work cooperatively, understand personal and physical differences among persons, make group decisions and improve communications skills.

Both the teacher and the child should learn from exploring and experimenting. The teacher-child relationship should be one of mutual respect. Children should not be competing or compared with one another.

Open school teachers should have respect for children, be keen observers and responsive listeners and be genuine learners and experimenters. Each teacher must be a skilled diagnostician of children's needs and stages of development, must provide viable options with multiple materials and resources, and must assist the child in making choices and evaluating his (her) activities. [Southeast Alternatives 1973, p. C-2]

Pupils at this school are not grouped by grade level. Instead, they choose a "family," which consists of two teachers and their aides and approximately fifty-five children aged from five to twelve. Each family is divided into two groups, ages five to eight and ages eight to twelve. Interaction between the older and the younger children is both encouraged and planned. The program that follows is one attempt to meet the philosophy stated above within this type of family organization.

The Integrated Curriculum

In the open school, the integrated curriculum best meets the philosophy discussed above. By an integrated curriculum we refer to one where children can integrate their own structures of knowledge by interacting with a wide variety of people and using a variety of materials, activities, and ideas. There is no one particular mathematics program, text series, or set of materials used to accomplish this. Piaget notes that children learn best through discovery based on real experience and that they need meaningful, concrete experience if they are to internalize or integrate mathematics with other subject areas. If we accept Piaget's premise, it means that during their elementary school years, many children are at a concrete operational level for most of their thought processes. Learning, then, needs to be concrete and relevant to each child's world, and children learn quickly when they see a reason or need. The following integrated activities are a few examples of ways to make learning relevant. Additional examples will be described in more detail later.

- Have the children build models, which will give them experience with space, substance, length, volume, area, weight, estimation, and measurement.

- Have the children make collages, which will give them experience in covering space, fitting shapes together, dividing, classifying, comparing different shapes, and matching patterns and sizes.
- Have the children stock and operate a "store," making the items to sell and carrying out all the necessary operations themselves, which will give them experience in figuring cost, profit, overhead, and prices.

Within this environment the teacher becomes a manager of time, space, and resources (both human and material), helping pupils find interests and asking probing questions. Mathematics concepts are experienced during the children's exploration of an interest. A particular concept or skill is taught by the teacher when a child discovers that the skill is needed in order to continue a project. Sometimes all the teacher must do to motivate the child to discover a solution is give a word of encouragement, ask a leading question, or make a comment. Because the child chooses the activities around which learning takes place, the teacher needs to provide an environment that offers each child a variety of opportunities for exploring and learning. The physical environment must stimulate interests, exploration, questioning, experimentation, and manipulation. It should provide for individual and group activities in both active and quiet places. It should provide space for areas that reflect possible activities and interests, for example, spaces for animals, games, plants, books, or puzzles. The following considerations have been suggested as important to keep in mind when designing room space (Hunter 1975, p. 6):

1. Use the noisier, more heavily traveled areas for the more active projects.
2. Use the lighter traffic zones for small and quiet spaces.
3. Save space for privacy, as well as room for whole-group meetings.
4. Have messy activities take place off the carpeting and near a sink.
5. Store a limited amount of frequently used supplies (scissors, rulers, glue, tape, crayons, stapler, assorted paper, and such) in a single location.
6. Store materials close to the area where the activity is most likely to occur (tapes and records near listening center).
7. Use furniture or dividers (macrame, tie-dyed sheets, yarn, string, etc., hanging from the ceiling) to define spaces.

Involving children in the decisions of organizing space often leads to stimulating integrated activities and can give them a feeling of ownership and pride in each area. An example of one such activity involved the planning and building of an indoor garden. The children decided that the garden should be placed in a position to capture the maximum amount of available sunlight, which led in turn to an investigation and discussion of angles and how angles are formed by rays. Pupils measured, timed, plotted, traced, and watched the areas of sun and shadow in the classroom. After collecting the data, they determined the shape of the garden (a parallelogram), the interior angles that would give maximum exposure to the sun, and the relationship between the dimensions of the window and the dimensions of the parallelogram. The children then built the structure—estimating, measuring, doing a cost analysis of materials, making a scale drawing, determining the volume of dirt needed, checking stress points, and determining the number of legs needed to support the structure safely. An excellent resource on the use of space in a classroom is *Arranging the Informal Classroom,* by Brenda S. Engel (1975[a]).

It is the role of the teacher to observe each child and interpret individual needs and readiness. Ways and techniques for developing an awareness of mathematics and methods for identifying the mathematical experiences of each child will be considered in detail later.

Ways in Which Integration Can Occur

There is no precise recipe or procedure that will guarantee an effective integrated curriculum. At times, all or most of the pupils may be involved in activities centering on one theme. At other times, individuals or small groups may be working on unrelated projects and activities. Either way, the teacher should continually be asking probing questions, answering questions, helping pupils find needed material and resources, helping them understand a particular concept or master a particular skill needed in order to continue the activity, and helping and encouraging them to develop or rekindle interest. Some detailed illustrations of integrated activities are described in the following two sections.

Whole-class project

Sometimes a particular topic or theme sparks the interest of the whole class. For a number of weeks all the students are engaged in activities and projects related to the theme. The theme itself can be a kind of umbrella under which a variety of investigations and experiences can be gathered and related. The challenges prepared by the Unified Sciences and Mathematics for Elementary Schools (USMES 1970–75) are good examples of possible whole-class projects.

Prehistoric or early people might provide the theme for a project by an entire class of eight-, nine-, ten-, and eleven-year-old children. How prehistoric people lived, how they structured their days, and what they valued could provide the focus for a myriad of activities that integrate language, mathematics, and the social sciences. The children might read; view film-strips; interview historians; collect, count, and classify fossils; and conduct research on what prehistoric times were like. An investigation of early counting systems for recording numbers of animals, for instance, might be undertaken. Calculations having to do with the average life spans of humans and different animals and the average land available for family or tribal units could be made. Children could draw pictures of people and animals in typical poses. When a grid is superimposed on these pictures, they serve as scale models for making enlarged cardboard models. In making the models, the children might try to solve the problem of using the cardboard so that the amount of waste is minimal. As they solve this problem, the topics of ratio and proportion, linear measurement, and area are experienced.

The cardboard models of people and animals could then be painted. If dry powder and water are used to make the paint, the children could calculate and test the effects of varying amounts of powder and water and note the resulting color. A chart such as the one in figure 10.1 might result and serve as a reference for other projects. The completed cardboard

Blue powder	Water	Color
15 g	250 ml	

Fig. 10.1

models can serve as props for a drama on prehistoric times written by the children. They could produce the drama for another class, for the entire school, or for a neighboring school. If appropriate equipment is available, the children could videotape the production. They could also design and make the costumes. Measuring, fitting, estimating, and computing would be a part of this experience. The production itself might be a culminating event for this whole-class project. In such an activity, which would take several weeks, it is difficult to distinguish the teaching and learning of mathematics from that of language arts and social science, for example. These disciplines have been integrated in a project that interests and makes sense to the child and also provides applications of mathematics that are meaningful.

Through a television show, film, book, story, news headline, new invention, prediction, or classroom interest center, the class might become interested in the idea of future life. Science fiction stories, films, TV shows, and computer simulations are rich and stimulating sources of ideas on what our future society might be like or what the children would consider an ideal culture to be like. This theme could have several ramifications. One might concern how and what we should preserve from our culture in a time capsule so that someone could learn about it a hundred years from now. A variation on this would be for each of two groups of children to design its own culture and to construct the artifacts to "preserve" it. The construction of artifacts would involve experiences with measuring, ratios, and proportions as well as the development of spatial relationships involved for finding the best fit in the capsule. Each group could work independently, prepare a time capsule, and hide it somewhere. Then each would provide the other with a map or scale drawing and directions for finding its capsule. When the capsules are discovered, the children could attempt to describe the culture from the artifacts they have found. A television or radio script might be written to announce the latest news concerning the progress and discoveries of each group. Individual children might also wish to keep logs or diaries of what is happening.

As a concluding activity, the children could discuss reasons for any discrepancies between what the designers intended and the way the discoverers interpreted or described the culture from the artifacts they found.

Again, a naturalistic approach to learning is evident in this theme. Instead of being compartmentalized, mathematics becomes an integrated part of the theme's activities. The design of future clothing, means of travel, and a communication system involve measurement, numerous calculations, and perhaps the solving of codes and ciphers. The coordinate system could be used and investigated in the construction of maps and guides to find time capsules. A future society might be expected to use computers and calculators on a daily basis. An investigation of computing devices, or the construction of a computer or logic machine, might be activities of the theme.

Another variation of a theme on the future could be a realistic view of our present environment. Based on present effects on the ecology, an analysis and projection of future society could be made. Will the future be nonexistent because of the destruction of "Spaceship Earth" (Fuller 1970)? If the planet Earth and its inhabitants are to survive, what changes need to be made? These questions provoke relevant and awareness-producing activities that can involve almost limitless mathematics at every level of sophistication. Experiments could be conducted to determine how much water is lost from a dripping faucet in one hour, how much food is wasted each day in the school cafeteria, or how much paper is wasted each day at the school. Charts such as the one in figure 10.2 could be made.

Children could attempt to find ways of recycling different materials, or they could collect paper, aluminum, or other materials for a recycling

Time elapsed	Amount of water dripped from faucet
15 minutes	
30 minutes	
45 minutes	
60 minutes	
75 minutes	
How much water would be wasted in a day?	
How much water would be wasted in a week?	
How much water would be wasted in a month?	

Fig. 10.2

center. This could lead to research to determine the amount of energy saved by recycling. For example, recycling aluminum saves 95 percent of the energy needed to produce aluminum from raw materials.

They could also try to determine the effect on all of life if one animal or one plant became extinct. They could construct a model environment and determine the effect on plant life of different types and amounts of air and water pollutants. Using the present rate of consumption and increase of costs, pupils could determine how long schools will be able to afford to use paper. They could devise a plan to sabotage ecologically the life-support system of "Spaceship Earth" or try to "design a workable spaceship capable of sustaining a maximum population of 200,000 people, plus essential plant, animal and insect life, for a 5000-year journey to another solar system" (McInnis 1972, p. 72). Possibilities abound for interesting and meaningful activities that involve mathematics.

Individual and small-group projects

Many times projects and interests of children are unrelated; several themes may be offered at once, or each child may be working independently on a project. If one child becomes interested in a particular topic and begins to work on a project, it could serve as motivation for other pupils also to work on the same or related activities. From television programs, field trips, other family members, books, films, or other experiences, they may become interested in such topics as energy, pollution, elections, or silversmithing. Children always seem to be interested in animals, particularly when some are kept in the classroom. Snakes seem to hold a special fascination for children. A child or a group of children very interested in snakes and reptiles could read stories about them and research their life-style, habitats, and diet. A pet snake may be kept in the classroom. Routine observations of food, water intake, or growth could be recorded, providing experiences in keeping records (fig. 10.3) and in making graphs (figs. 10.4 and 10.5) based on the data kept.

Snake Chart

Name of Snake _______________________________

Date of Birth (if known) _______________________

Date	Length	Weight	Temperature in cage	Temperature outside cage	Food intake

Fig. 10.3

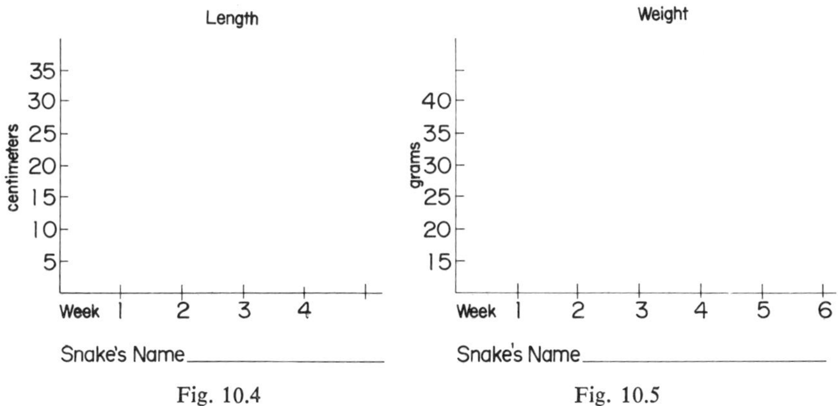

Fig. 10.4   Fig. 10.5

A map might be made showing where different kinds of snakes are likely to be found. This might lead to a classification of poisonous and non-poisonous snakes and an investigation of venom and antivenins. A visit to a zoo or university where snakes are on display could be arranged. The child might also write a story about a snake's perspective of people.

Some commercial or teacher-made games are usually found in every classroom. Games like Monopoly, chess, checkers, Stratego, and Battleship always seem to attract and interest children. Through this interest in games, pupils could become interested in inventing and designing their own games. Those who are dissatisfied with an existing game could be challenged to improve it or to make a better game. In this way, games might become the topic for one small-group project. The group might develop games related to special days—a Groundhog Day counting game or a Halloween multiplication game, perhaps. Rules are agreed on, scoring is set, and the game board is constructed; then the game is played. After playing the game a few times and correcting any problems in directions or rules, the puplis may decide to have other classmates participate in a room-wide or school-wide tournament. They may decide that their game is better than many commercial games and may want to try to market it. They could produce and sell the games themselves, or they could try to sell the game to an existing company. Games might be made by older children to help younger children practice counting or basic-skills concepts. In this way the game makers use their mathematical skills in new and creative ways, and the game players have a game designed especially for them.

Another small-group project might focus on athletic games or events. The history of gymnastics, track, baseball, or football might be investigated and the evolution of present-day rules traced. The impact on sports of the change to the metric system might be researched. Children might read daily or weekend newspapers and record scores of local school teams or national teams. Records of well-known players could be kept and averages calculated for individual players or teams, as shown in figure 10.6. A simulation, which could be computer based or non–computer based, might be developed and tested. Children who understand a particular sport might try to teach it to those who are unfamiliar with it.

Names of players	Yards gained in each game								Average for season
	1	2	3	4	5	6	7	8	
			Team average						

Fig. 10.6

Projects might last a week or two or continue throughout the year. This depends on the interests and needs of the children who are involved.

Evaluation and Record Keeping

The method of evaluation used in a classroom or a school can set its educational tone. Tests that emphasize competition in testing isolated skills and that are used to sort, rank, and label children tend to make teachers look at students' differences as deficiencies rather than as strengths and unique qualities (Devaney 1974, p. 148). Devaney has noted the uniqueness of individuals and their achievement (p. 147):

> Learning is different for each child not only in pace but style and content. Learning should *not* be measured by instruments which count how many separate, standard concepts and skills the child can demonstrate. It should be assessed by procedures which sense how each child is gathering intellectual power.

Evaluation should be a method of determining what children *can* do in realistic settings rather than what they cannot do. Two important points emerge: evaluations need to furnish a method of finding the children's strengths and experiences, and they need to demonstrate how children use the skills and knowledge they have. This can be done by keeping a record or file containing examples of pupils' work and a list of their experiences. But if a record is to be kept of a child's mathematics experiences, the teacher and pupil need to be aware of the mathematics involved in the activities and projects.

When an integrated curriculum is used, one of the problems is making teachers, pupils, and parents realize—and feel comfortable about the fact—that children *are* having mathematical experiences. It seems that so many years have been spent in teaching mathematics as a set of isolated skills unrelated to practical situations that people have come to view the isolated skills as all that mathematics is. Therefore, teachers and pupils need experience in identifying the mathematics that is involved in most activities. Once an awareness is developed, mathematics can be seen in most activities.

One method of developing this awareness is to spend five or ten minutes each day identifying the mathematics involved in a project or activity that does not appear to be a mathematics activity. For example, making a collage could involve the concepts of space, area, volume, matching, sorting, combining different shapes, measuring, estimating, problem solving, counting, balance, symmetry, dividing, partitioning, fractions, and other topics. The group should try to explain how the mathematics is involved.

Another method of developing awareness and identifying the mathematics the child has used is through observation. The observer (the teacher or other adult) notes what each child or group of children is doing. After taking notes on the activity, the children involved, and what happened, the observer compiles a chart identifying the mathematics concepts involved (fig. 10.7). This provides a record of the pupils' activities and their mathematical experiences.

Observation Chart

Children involved	Activity	Materials	Math concepts involved
Ruby George Ramone	Math game— *Metric 21*	Deck of *Metric 21* cards, centimeter ruler, chips	estimating, metric measuring, computing, probability, problem solving, sequencing, counting, comparing
Maria	Building an apartment building about two meters tall (it fell over; so she concluded that top pieces should be placed at the bottom as reinforcement)	scissors, ruler, cardboard, paste, tape	estimating, problem solving, shapes, area, volume, surface area, joining, sequencing, counting, computing, angles, edges, faces, 3-D construction, strength of structures, balance

Fig. 10.7

The mathematics checklist shown in figure 10.8 can be used to document children's involvement in, and their ability to use, mathematics in realistic settings. Along the side are the mathematics objectives considered essential for all pupils. Across the top are some of the activities in which they might be involved. Additional or alternative activities would be written in for individual children. Each objective that is used during the activity is checked. This type of checklist gives the teacher an overview of the children's strengths and experiences in mathematics.

Another method of identifying the mathematics a child is using is through a technique used by Brenda Engel (1975[b]). An observer identifies the mathematics that each child uses during a five-minute period. After the observation, the information is coded on graph paper with the children's names down the side and with each square representing a five-minute block of time. Each mathematics activity is assigned a color, and each five-minute block is color coded for each child. Symbol codes such as / / /, indicating that a pupil is "unconcentrated," or X X X, indicating that

Essentials for All Students	Monopoly	Straw sculpture	Store	Reading and using a recipe	Making "for sale" poster for store	Making and using a top	Origami
Concrete experience with number concept	✓	✓	✓	✓	✓	✓	✓
Concrete experience with place value	✓		✓				
Concrete experience with combining groups (addition)	✓	✓	✓	✓		✓	✓
Concrete experience with breaking groups into smaller groups	✓	✓	✓			✓	✓
Experience with a calculator			✓				
Use of arithmetic operations and basic facts	✓	✓	✓	✓	✓	✓	✓
Experience with estimating solutions or outcomes	✓	✓	✓		✓	✓	✓
Experience with estimating measures		✓	✓	✓	✓	✓	✓
Concrete experience with measuring		✓	✓	✓	✓	✓	✓
Concrete experience with space, shape, solids, volume, weight, area, and symmetry		✓	✓	✓	✓	✓	✓
Experience with a situation that needs a solution, the need to make decisions (problem solving)	✓	✓	✓	✓	✓	✓	✓
Concrete experience with basic fractional parts and division into parts			✓	✓	✓	✓	✓
Opportunity for positive feelings about mathematics	✓	✓	✓	✓	✓	✓	✓
Possibility for open-ended approach, a chance for inventiveness and creativity		✓	✓		✓	✓	
Opportunity for satisfaction from accomplishing an activity or project	✓	✓	✓	✓	✓	✓	✓

Fig. 10.8. Mathematics checklist (adapted from Fehlen [1975])

the child is engaged in conversation, can be used. This gives an overall picture of the activities with which pupils are involved.

A mathematics monitoring chart (fig. 10.9) is a checklist of objectives that the teacher can use to get an overview of the child's strengths and experiences in mathematics. It can be filled in by the teacher or by the teacher and pupil together. Some teachers find this type of chart a good summary sheet to use for parent-teacher conferences.

The mathematics monitoring chart can also be used to develop mathematics awareness and as a source of ideas for activities for each objective. The teacher, either individually, with other teachers, or with a group of pupils, can compile a list of materials and activities that give experiences involving each objective. For example, the list of experiences and materials

Mathematics Monitoring Chart

Pupil's Name _________________________ Date of Birth_________________

(day, month, year)

1.	The child feels good about mathematics.	rarely	sometimes	frequently
2.	The child has concrete experience with number concepts.	rarely	sometimes	frequently
3.	The child understands number concepts.	rarely	sometimes	frequently
4.	The child has concrete experience with place value.	rarely	sometimes	frequently
5.	The child understands place value.	rarely	sometimes	frequently
6.	The child has concrete experience in combining groups.	rarely	sometimes	frequently
7.	The child realizes that addition means putting together or combining groups.	rarely	sometimes	frequently
8.	The child realizes that subtraction means breaking groups into smaller groups, the reverse of addition.	rarely	sometimes	frequently
.				
.				
.				
23.	The child has a good self-concept as a problem solver.	rarely	sometimes	frequently
24.	The child knows how to find needed information.	rarely	sometimes	frequently
25.	The child seeks help or information in trying to solve the problem.	rarely	sometimes	frequently

| 26. | What type of help or information does the child seek? |

| 27. | Where does the child go for help or information? Who does the child ask for help or information? |

| 28. | Other mathematical experiences not on the chart. |

| 29. | Other mathematical behaviors noted. |

Fig. 10.9. Adapted from Marcy Open School (1975)

in figure 10.10 would include examples for the second objective listed on the mathematics monitoring chart in figure 10.9.

2. The child has concrete experience with number concepts.

Experiences	*Materials*
walking	stones
making playing cards	leaves
taking class attendance	people
dancing	tickets
doing needlepoint	books
sorting leaves	seeds
knitting	windows
weaving	nails
crocheting	pencils
following directions	playing cards
paper folding	paper clips
stamp, coin collecting	buttons
finding a page	doors
pacing a distance	raisins
building with blocks	shells
comparing	blocks
sorting	pinecones
counting	flowers
matching	paints
ordering	shoes
	apples
	sounds

Fig. 10.10

The methods described above can help teachers find the individual child's strengths and experiences. These methods are being used in some open schools. It is necessary to note that teachers, like students, are also individuals and should use the record-keeping methods in ways that best fit their individuality.

Conclusion

Although the open school is the option or alternative identified here, the integrated curriculum could be a viable method for other alternative schools. A particular kind of alternative school is chosen to fit the needs and learning style of an individual child. The "best" alternative varies for each child. Our intent here is not to say that one alternative style is better than another, but rather to point out that alternatives are necessary and that regardless of the alternative model used, learning needs to be relevant and built on the strengths and experiences of each child. The integrated curriculum is just one method of accomplishing this goal. As described, it attempts to take into account the individuality of each child and tries to emphasize a meaningful, concrete holistic approach to learning in the tradition of the open school movement. The authors realize that as described, it is both realistic and idealistic. This essay is meant to illustrate some of the practices and philosophy that underlie individualizing the teaching and learning process using the integrated curriculum.

BIBLIOGRAPHY

Anderson, Joyce A., Joan E. Fehlen, and Mary Lou Hartley. "The 3 R's in an Integrated Curriculum." *Mathematics Journal* [Minnesota Council of Teachers of Mathematics], Spring 1975.

Barth, R. S. *Open Education and the American Schools.* New York: Agathon Press, 1972.

Devaney, Kathleen. *Developing Open Education in America.* Washington, D.C.: National Association for the Education of Young Children, 1974.

Engel, Brenda S. (a). *Arranging the Informal Classroom.* Grand Forks, N.D.: University of North Dakota Press, 1975.

——— (b). *A Handbook on Documentation.* Grand Forks, N.D.: University of North Dakota Press, 1975.

Fehlen, Joan E. "Math in the Integrated Curriculum." Mimeographed. Minneapolis: Southeast Alternatives Project, Minneapolis Public Schools, 1975.

Fuller, R. Buckminster. *Operating Manual for Spaceship Earth.* New York: Pocket Books, 1970.

Hunter, Peggy. *How to Put It All Together: A Handbook on Organizing Space in the Informal Classroom.* Minneapolis: University of Minnesota and Minneapolis Public Schools Teacher Center, 1975.

McInnis, Noel. *You Are an Environment.* Chicago: Center for Curriculum Design, 1972.

Marcy Open School, Minneapolis Public Schools. "Mathematics Monitoring Chart." Mimeographed. Minneapolis: The School, 1975.

North Central Association of Colleges and Schools. *Policies and Standards for the Approval of Optional Schools and Special Function Schools.* Chicago: The Association, 1974–75.

Silberman, Charles E., ed. *The Open Classroom Reader.* New York: Random House, 1973.

Southeast Alternatives Project, Minneapolis Public Schools. "Southeast Alternatives 1971–1976 Plan." Mimeographed. Minneapolis: Minneapolis Public Schools, 1973.

Unified Sciences and Mathematics for Elementary Schools (USMES). Challenges series. Newton, Mass.: Education Development Center, 1970–75.

11

The Teacher-centered Mathematics Classroom

Gerald R. Rising

Stephen I. Brown

Lawrence N. Meyerson

*T*wo questions may immediately occur to the reader: (1) Why should an essay on the teacher-centered or self-contained classroom appear in a collection of essays about alternative forms of organization? (2) Why should such an essay be proposed at all? Before turning to our central topics, we respond to these questions.

Why should this essay appear in this collection? Over the past decade, the literature of education in general and mathematics education in particular has been saturated with articles on a wide variety of forms of classroom organization: individualization of many types, laboratories, groups of different sizes, classrooms without walls, and so on and on. Moreover, the contemporary thrust of many teacher-education programs is directly associated with one or more of these panaceas. We have visited schools in which one or another of these organizational forms is adopted by *all* teachers for *all* mathematics instruction. To some teachers in some schools and to an increasing number of teacher educators, then, the teacher-centered classroom is indeed not only an alternative but even an exceptional organizational procedure.

An early draft of this paper was presented at a faculty-student seminar at SUNY at Buffalo. Helpful suggestions were made by LeRoy Callahan, Don Cockerill, Bryan Gillette, Stephen Hicks, George Schena, Paul Schwiegerling, and Ashwin Shukla.

"

The second question is more important: Why should such an essay be prepared at all? Most of the articles we have read about new styles of classroom organization use the self-contained classroom as a scapegoat. If all the things said against teacher-centered organization were true, then surely a defense of that style of teaching would be a disservice to mathematics education. Obviously, we do not agree. We suggest that the basic problem is that teacher-centered instruction has been most unfortunately equated with bad teaching.

We shall now describe and justify the teacher-centered mathematics classroom. First, we shall describe the organization of such a classroom at its best. Then we shall respond to the criticisms of this method and the commendations of alternatives as they apply to our description.

The Teacher-centered Mathematics Classroom

In this section we provide a positive view of the teacher-centered mathematics classroom. We describe what it *can* be in the hands of a master teacher. To defend ourselves against the charge of wearing rose-colored glasses, we also offer a series of vignettes that will illustrate and extend our descriptions. Each of these vignettes represents an episode one of us has observed. Together with the descriptions of this type of classroom, they provide a peephole into the superior teacher-centered mathematics classroom.

We use *teacher-centered,* not in the sense of the teacher being the center of activity, but rather of his or her being the control center, the center of organization and planning. In particular, the teacher-centered classroom is not simply or primarily a lecture hall. (This does not imply that lecturing is always inappropriate, even in school mathematics. This point will be discussed later.) There, and in many of the other alternative forms of organization, teachers are restricted to a single mode of presentation. Here, teachers provide the mode that in their judgment best fits their immediate and long-term objectives. The specific activity may be a game, a test, individualized or group work, a lecture, a tutorial, work with concrete objects, a field trip, a film, a student report, a discussion, board work, team teaching with students or colleagues. Only the teacher's own creativity, judgment, and style provide limitations.

Item 1

The teacher has performed a magic trick with a deck of cards in an elementary algebra class. The basis of the "trick" lies in algebraic relationships. He repeats the trick, calling on different students to participate in minor roles. Then he asks for volunteers who think they have mastered the steps of the trick. To each of these students he gives a deck of cards and assigns an

audience of more naive classmates. The teacher monitors the group activity, correcting procedural errors and encouraging others to try performing the trick. Soon he identifies the students who have mastered the technique and regroups them to search out the underlying mathematical bases for the consistent outcome of the trick. By the end of the class period, several of these students have discovered various algebraic bases for the trick. These students are urged to consider modifications of the trick for homework. Many other students are near the solution; they are encouraged to complete it. A few have just mastered the techniques; they are asked to try the trick on friends or family before trying to explore the basis. In a subsequent class the algebraic relationships extracted from this activity will be employed and extended. In addition, the students will be made aware that they had been given different homework assignments based on their differing conceptions of what was in-
· volved. They will then explore the appropriateness of these assignments for each of their different learning styles. The students will then be given some responsibility for defining the type of homework assignments appropriate to their style of learning.

Item 2

The third-grade teacher has written on the chalkboard the sequence of counting numbers 1 through 30. He has asked his students to represent as many of the numbers as possible between 1 and 100 as sums of two or more adjacent terms in the sequence. To present this in terms that third-graders can understand, he has given these examples:

$$15 = 4 + 5 + 6 \text{ and } 15 = 7 + 8$$
$$75 = 13 + 14 + 15 + 16 + 17$$

Students work in pairs. In monitoring the work, the teacher encourages systematic search—what the students call "looking for shortcuts" or "saving time." After about ten minutes he stops the small-group work to allow one group to report its discovery on how to find the multiples of three: "To get the three times table, you take three adjacent numbers. They will always equal three times the middle number." Several others have discovered this rule. One suggests why the rule works. But as another student offers to extend the rule the teacher has her whisper her secret to him. He then tells the class, "Sally has a good way to extend the rule to other multiples. I won't have her announce her discovery because that would take away your fun." And then the students return to their task.

In the teacher-centered classroom, the teacher is the central figure in planning and interpreting curricular materials; in interacting with, and—more important—developing interactions among, students; in orchestrating classroom activities; in identifying and responding to individual, small-group, and whole-class needs; and in maintaining a positive classroom atmosphere, an atmosphere where "fun" seldom displaces learning but may indeed contribute to it. What distinguishes this mode of instruction at its best from most others is not that teachers take major responsibility for deciding on the pattern of teaching, but that they make that decision not just once at the beginning of the year but many times each day.

Item 3

An eleventh-grade class is playing the game Clues (not to be confused with the commercial board game, Clue). The students are divided into two teams whose members seek to solve a complex problem by a complicated process of noting contradictions among statements. In the process they must avoid traps set by false statements. Points are scored or lost at each turn, but the excitement builds because the final solution usually determines the winning team. This usually blasé group of teenagers is as excited as a group of fifth graders. The teacher's role in the game is only to expedite and referee.

Item 4

The ninth-grade mathematics and English teachers have cosponsored a limerick contest in which the students are to add to this opening line:

There once was a number named Zero . . .

Intellectual growth and attainment are intimately tied to communication. This is the postulate that is fundamental to the well-managed teacher-centered classroom. It is the inability of some other classroom organizational modes to incorporate this postulate into the management system that makes them as sterile as some of the similarly restricted geometries. Communication here takes two forms. First and quite obvious is the communication of ideas *to* the student not only by the teacher but also by other students—and even by the student himself. Second, and of greater importance, is the communication of ideas *from* the student. The importance of communication in this direction is that first of all the teacher can thereby diagnose the nature of difficulties students are having; second, fellow students may be in a position to learn from language (no matter how crude) that is more akin to theirs than the teacher's is; and third, the teacher may be influenced by a student's insight to redirect an investigation that had been perceived in a more limited light. Finally, if students are successful in *explaining* an idea (and not just in answering a question), then we have some confidence that they are beginning to understand the idea.

Item 5

The sixth-grade mathematics class has generated a lengthy list of statements about rational numbers. These statements fill several chalkboards. They range from

> "The top number in a fraction is called
> the numerator.—John B."

to

> "When you make the denominator of
> a fraction bigger, you make the frac-
> tion smaller.—Billy Joe P."

and

> "To change a decimal to a fraction,
> put it over 10 or 100.—Sarah M."

(In this class the teacher wisely plays down the number-numeral distinction between rational number and fraction so important to the formalist but so tangential to mathematical progress at this age.) The students have now run out of statements, and the teacher engages them in a discussion of their list: "Are any of the statements wrong? Can you show the author why his or her theorem is incorrect? Can another student suggest a way to correct the statement? Will the author accept that suggestion? Do any statements say essentially the same thing so that we can shorten our list? Sarah, will you show me by example how your theorem works? . . . Now on the basis of our observations, do we have any new statements?"

What is missing in so much poor teaching—teacher centered or not—is the responsibility associated with the role of the teacher. So many thousands of dull, routine classrooms across the country are staffed by pedestrian teachers who either fail to recognize their responsibility or fail to fulfill it. Their classes have degenerated into 38 minutes of reviewing yesterday's homework and one minute of giving the new assignment. But homework as well as the entire spectrum of teaching can be handled in a creative and intellectually stimulating manner.

Item 6

During the middle of a calculus lesson on the derivative, one of the students remarks, "If $y = x^2$, then when $x = 4$, $y = 16$. We learned yesterday that the derivative of a constant is zero, and therefore y' should be 0 when $x = 4$." Instead of answering the question directly, the teacher asks for opinions of the student's remark. She soon realizes that there is a great diversity of opinion and of evidence for the opinion. She therefore gives the class the following homework assignment:

Given the equation $y = x^2$:
1. Supply two arguments to support the claim that $y' = 8$ when $x = 4$.
2. Supply two arguments to support the claim that $y' = 0$ when $x = 4$.
3. Which arguments are the strongest?
4. Which do you believe to be the correct answer and why?

The next day the teacher collects the assignments and prepares a full lecture built around an analysis of the responses, giving more credit to well-supported wrong arguments than to poorly reasoned correct ones.

To give that assignment, it was necessary for the teacher to take advantage of the moment, to appreciate that a diversity of opinion is frequently a motivation for serious thought, to realize that students take a huge step in the direction of intellectual maturity when they try to argue for a position with which they disagree.

If the way in which this homework was handled was a nonroutine and perhaps one-shot strategy, there are others that are creative and stimulating to the student and yet provide a degree of security.

Item 7

> As students enter the geometry classroom and unpack their books, several mark on the chalkboard the numbers of the homework exercises with which they had difficulty. Other students who were able to solve those problems write out what they believe were the key steps in their solutions. The teacher makes use of these solutions to lead into the new lesson. Later in the class period, the teacher is able to use the offered solutions to limit the discussion of homework to about ten minutes. The teacher has carefully developed this responsible procedure, and the students are now comfortable with it.

Teacher-centered organization does mean routine when routine contributes to the progress of the class, as in the last item. Most teachers know the value routine has for giving students a feeling of security in the classroom. Established routines calm down the most difficult classes. But fewer teachers recognize the conservative nature of students, which makes breaking their routine a difficult process. Once students have their wheels in a rut, it is difficult to get them to steer out of that rut, even when the routines are extremely boring or oppressive. Too many teachers have tried new techniques only to reject them without overcoming this initial reaction of their students.

The master teacher orchestrating a teacher-centered classroom seeks a balance between (1) routines that move the class activity forward efficiently and provide students with a measure of security and (2) novel activities that change the pace, enliven the class, and encourage creative student activity. (Researchers in education are frequently warned to beware of the Hawthorne effect: that the novelty itself of the experimental design might account for any success of a method. Practitioners, however, have not yet learned that here is something to capitalize on—if novelty breeds success, then instead of controlling for it, introduce it intentionally.) Classes run only with set routines tend over a period of time to run downhill to boredom; classes run only with novel activities tend to leave students with organizational problems. As is so often true, a careful synthesis of these two styles, designed to draw out the best of each and to play down their disadvantages, is the answer. And the good teacher knows how to adjust the balance to compensate for difficult times, such as the day of a major athletic contest or the day before a holiday.

Item 8

> The eighth-grade class is broken up into groups of four students, each group working on difficult, nontraditional problems. For example, the following problem is being attacked in one group:
>
> In the figure below, a 3×5 rectangle is divided into unit squares. The diagonal of the rectangle passes through seven of these squares. Try to determine a rela-

tionship between the width (w), the length (l), and the number of squares through which the diagonal will pass (s) for any rectangle with whole-number sides.

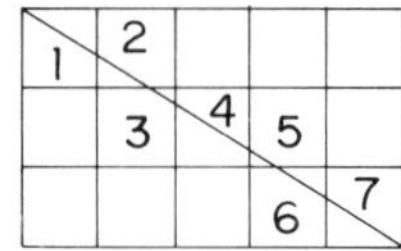

The students in this group are using centimeter graph paper to help them generate data. They have already noted that for any rectangle having a width of one unit, the diagonal passes through all the squares. They are now trying rectangles having a width of two units.

Not only did the teacher in this class decide to select at this moment nontraditional problems having the potential to develop mathematical thought in the absence of considerable machinery and jargon, but she also had to decide that *for this class at this time* it was appropriate to have them work in small groups and on *different* problems in each of the groups. Furthermore, she had to establish criteria for forming the small groups. The number of variables that the master teacher must juggle at any moment in order to account for characteristics of both individuals and the entire class is astounding. To make a wise decision is not easy, but imagine how much more significant in the lives of students decisions of this sort are, compared with monolithic decisions made on the first day to commit oneself to a program having few points of choice.

Item 9

In a ninth-grade class for students who have had little or no success with mathematics, the teacher has assigned twenty minutes each Friday for work on individual progress worksheets that focus almost exclusively on routine calculation. The students keep records of their personal progress and move ahead in the sequence only as they are successful at each stage. The teacher uses this time to observe individual students carefully in order to note special difficulties. He will respond to difficulties involving the total group in total-group settings, and to difficulties involving smaller groups in smaller settings; he will respond to an individual student's difficulties on a one-to-one basis.

Item 10

A twelfth-year student has been absent for three weeks. The teacher provides him with a programmed-instruction algebra text that covers the material he missed and works out a schedule that allows him to make up work and receive additional assistance from classmates and the teacher.

In summary, the teacher-centered mathematics classroom is a classroom carefully organized by the teacher to bring out the best in students (and teacher) and to optimize their mathematical development. To do this, teachers must draw on their own knowledge, style, and experience and adapt these personal resources to fit both immediate and long-range goals. From these goals, teachers develop their own specific classroom activities. They feel no limitation from a specific method but rather adopt without hesitation whatever technique they believe will work best for the given objectives, for the given students, and for themselves.

The purpose of the vignettes in this section is not merely to describe "cute" experiences inside the mathematics classroom, but rather to explore new ideas in mathematics teaching around the concept of the teacher-centered classroom. The specific items are merely illustrative. How do these items bring out these methods? This question can be approached by asking a somewhat different question: What do most of these items have in common?

A quick superficial search would indicate that these items are simply different pedagogical procedures for getting at specific concepts. This is only a partial answer. A deeper exploration yields teaching methods that can be *generalized* to cover many classroom situations. For example:

1. The vignettes are incidents in which the teacher must use more than one teaching method. Consider the card trick in item 1 as an example. Here the teacher must deal with student subgroups at different levels of understanding as well as with the group as a whole. He must concern himself with those students who improvise on the trick mathematically as well as with those who need assistance in understanding basic parts of the trick itself.

2. The incidents force students to ask questions. Consider item 6, in which the students generate a list of statements about rational numbers. From these "data" the students are asked to *pose* problems about what *they* have observed. In this way the mathematics developed is truly a function of student perceptions, not just answers to specific questions asked by a text or trumped-up "discovery exercises."

3. The incidents suggest that teachers should use students' work in their presentations. Consider, for example, the geometry class in which students wrote on the chalkboard their own solutions to problems (item 7). Instead of solely evaluating the solutions as right or wrong, the teacher uses the solutions (correct or incorrect) as a partial introduction to a new lesson. Instead of the homework being an ending point, it becomes a starting point for new explorations.

4. The incidents suggest an extension of both instruction and curriculum. A major concern of many teachers is the effort it takes to attempt to respond to all ability groups within the classroom. We

extend this concern to suggest that students come in more interesting varieties than low, average, and above average. These incidents demonstrate that students of differing abilities and orientations can "brainstorm" together, but they also can profit from extending their individual perceptions on their own. In the card trick in item 1, one student, after a brief discussion, makes it her own by attempting to come up with a formal proof; another creates a variation of the trick that seems to work despite the fact that he cannot prove it. In this manner there is something for everyone.

Facing the Alternatives

Implicit in earlier statements are several answers to the widespread and severe criticism of teacher-centered instruction:

1. Much of the criticism has really been directed at poor teaching and not at this mode of classroom organization.
2. Teacher-centered instruction is not restricted to the lecture method.
3. Teacher-centered instruction easily adopts—or more often adapts— good parts of any alternative teaching procedures. At the same time it is not chained to those routines.

Let us explore these answers further.

Poor teaching

All would agree that poor teaching is unacceptable in our schools. Students (and other teachers as well) should be protected from it. It poisons the educational system and demands direct personal, supervisory, and administrative response. Some of us recall with deep resentment, for example, teachers who were assigned lighter schedules or alternative nonteaching tasks like book distributor (sometimes even counselor or principal!) in lieu of teaching because they were failures in the classroom. Many of today's panaceas still dodge the issue by changing the teacher's role into that of clerk. Surely this is not an answer. By doing this, we not only avoid the issue but also block any attempt at solution.

Consider in this regard, for example, the progress toward giving tenure to teachers in the management type of classroom. They are adequately "managing" the essentially self-managing system. They monitor the routines, see to it that the worksheets are properly filed and refiled, check student records, and respond to questions from individual students. In some of these classrooms that we have observed, better than 90 percent of the "teacher's" role is clerical. And the other 10 percent is restricted to repetitious one-to-one interactions. Are these people teaching in any real sense of that word? Can they teach? We suggest that the answer to the first question is no, and that neither we nor anyone else, including the person evaluated, know on the basis of this evidence the answer to the second question. Going further, we suggest that whatever teaching ability these persons may have had at one time has atrophied under these conditions. Granting a person who has had only this limited and sterile experience and who has demonstrated virtually no teaching competence a lifetime appointment as our colleague is an act of unacceptable risk to the future improvement of even the maintenance of pedagogical standards. And yet this type of system provides exactly that kind of basis for evaluation. It is in this sense that we see some of the alternatives, particularly programmed worksheet individualization, as *blocking* solutions to educational problems.

Some would say that teachers are so bad that this change of role is warranted. These critics are the promoters of "teacherproof" methods and materials. We agree that too much of teaching is pedestrian, but we disagree directly with their response. The best response, we claim, is to attack the problem by positive means.

Teacher-centered instruction and the lecture method

Equating teacher-centered instruction with the lecture method smacks of a debater's ploy. The association is indeed unfortunate because of the many sterile examples of lectures that every college student can recall. Where this technique is employed exclusively or even extensively in school mathematics classes, we decry it. It is our observation, however, that it is not a widely used teaching style in the schools.

In fact, it is worth turning aside for a moment to say a few words in support of a good lecture appropriately used. In this regard consider a single, often forgotten aspect of teaching that can occur in almost its pure form in a lecture: the teacher as a model of a mind at work. The story is told of a lecture by Albert Einstein in which, as he completed a proof that brought order to particularly chaotic and seemingly conflicting evidence, tears came to his eyes—an extreme example that focuses on emotions, perhaps, but any reasonably good teacher has had the experience of similarly touching the heart and mind of a student. Youngsters need models not only of organizational engineers but of people committed to a body of knowledge and willing to display emotional and intellectual strengths as well as human foibles as they attempt to make sense out of their world.

Teacher-centered instruction as adaptive instruction

To the well-known line
I do and I understand
we add the words "only superficially" and the additional line
I discuss and I make my own.

The work of Stanley Erlwanger, Herbert Ginsburg, and others has shown that students working independently do make significant progress—to the credit of the programs it would seem—but they also demonstrate that progress can be, and indeed often is, in the wrong direction. Examples of this abound. We quote only one, communicated to us by Robert Davis. A student who had successfully completed an individualized unit on decimal fractions responded as follows to a series of exercises:

$$.2 + .3 = \underline{.5}$$
$$2. + 3. = \underline{5.}$$
$$2. + .3 = \underline{.5.}$$

Certainly the student understood something. He learned. His pattern (too often the only type of clue that individualized programs use for instructing) is indeed creative, but unfortunately it is not mathematics. What is missing in this individualized instruction is interaction with other students and with an insightful teacher. We do not question either the efficiency or the tight focus of individualized programs. What we do question is the absence of the opportunity to test one's ideas against those of one's peers in the intellectual marketplace, an aspect of learning that is of paramount importance.

In fact, even small student groups do not entirely respond to this concern. They at least provide interaction, but anyone who has observed such groups closely is bound to be struck by the need for occasional intervention from a person with a broader knowledge of the subject and a deeper understanding of what lies ahead in the curriculum. That intervention, we claim, cannot be supplied merely by a teacher hovering somewhere in the background. Granted that no matter what the organizational mode may be, some problems of this type will arise; still, it is our observation that these problems tend to increase exponentially over the time an exclusive mode is employed.

Note that the concept of teacher-centered instruction as adaptive responds to most of the claims of the alternative modes as well. Let us consider some examples. Certainly *individualization* motivates some students by letting them work at their own pace. Recognizing this, teachers should use individualized materials when this and other features of the program best serve their goals. Review and reinforcement activities, for instance, would be prime candidates for individualization. Certainly the *laboratory method* provides opportunities for participation in data gathering and contact with concrete and real-world materials. Here again wise teachers are sensitive to these values. They could use this method to generate problems and to set the stage for conceptual development. They could use it equally well to provide applications for following up theoretical work. Certainly *working in small groups* encourages students to make progress with more difficult problems. The group interaction carries students over rough spots that would have made them quit if working individually. Teachers can use this method to promote intellectual growth and even to develop concepts for follow-up organization in full-class settings.

Just as these methods provide specific boosts to the total program, however, so are they by no means interchangeable. To see this in a striking way, merely reread the last paragraph interchanging the words *individualization, laboratory method,* and *small groups* in any of the other possible permutations. Neither their advantages nor their applications shift comfortably. If they are not interchangeable for these specific activities, surely they are unsuitable exclusively for any total program. Q.E.D.

A Concluding Note

Because of space restrictions, we cannot extend our analysis of the issues to deeper levels. We believe, for example, that some of the slogans cited in support of classroom reorganization do not stand up to close scrutiny. Among these are "Children should work at their own pace" (individualization) and "Make learning concrete and not abstract" (laboratories). We also believe that the empirical tradition taken over from the physical scientists by researchers in mathematics education is leading away from reasonable, humanistic solutions to teaching problems. We need more careful, strictly philosophical, analysis of (for example) the various meanings of "to know" and "to teach" before we can barge ahead with behavioral objectives and paper-and-pencil tests designed to prove one teaching regimen better than another. We also need to learn how to observe and interpret the intuitions of master teachers whose wisdom may not be reflected in chi squares!

Even without this deeper analysis (which we are currently preparing for publication), we believe that the case we have presented places the teacher-centered classroom in a fairer perspective. We now invite the reader to reconsider two fundamental questions: Why is the teacher-centered classroom being consigned to the organizational junk heap? And what in the current climate is leading the mathematics-education community to leap on an unending succession of bandwagons?

12

Implications of Research for Instruction in Self-paced Mathematics Classrooms

Harold L. Schoen

*I*n the 1960s and 1970s there has been renewed emphasis on the responsibility of schools to meet the needs of individual students. Not since the peak of the progressive education movement have educators focused so directly on the individual. This phenomenon is reflected in the professional education literature of recent years. For example, in 1971 *Education Index* listed 124 articles on individualized instruction; the average number was about 35 a year during the decade of the sixties. In contrast, only about 4 or 5 (and often fewer) articles on individualized instruction appeared each year in the forties and the fifties (Kozak 1974).

Few educators would deny that the schools *should* be meeting the needs of the individual as much as possible. Any disagreement on this issue concerns the means by which these individual needs can best be met. Many approaches have been attempted. Good teachers in conventional classrooms have always used some effective techniques for individualizing instruction. Ability grouping has been and is used in many schools as a step toward individualization. The concept of the free school in recent years and the progressive education movement of the early twentieth century are two alternative instructional approaches that were designed to meet the needs of individual students. In the 1920s, the modularized, self-paced, continuous-progress, systems approach to individualized instruction gained widespread popularity, only to virtually disappear in the 1930s (Grittner 1971). In recent years we have witnessed a renewed interest in updated versions of this approach that will be referred to here as self-paced instruction (SPI).

SPI involves student self-pacing using a sequence of learning packets or units. Many SPI programs involve a variety of media, but for the most part students learn independently from textbooks and worksheets. Pretests and posttests are included with each learning packet, and passing one or both of these is a prerequisite to progressing to the next packet. The teacher's role is that of manager, record keeper, individual tutor, and sometimes curriculum developer. Unlike the progressive and free-school approaches, SPI does not provide an alternative curriculum. Rather, it provides a detailed, highly structured, carefully sequenced guide that can lead students to learn all the components of the traditional curriculum more or less independently. Most SPI programs also include a means whereby teachers can use data to diagnose problems of individual learners, prescribe appropriate instruction, and maintain a detailed record of each student's progress through the curriculum.

The case for SPI has been stated often and eloquently, and some equally eloquent critics have emerged; in the end, empirical evidence must determine whether or not SPI is effective. The main issue is, "Does SPI have educational benefits, and do they outweigh any liabilities?" Two brief reviews of research on SPI mathematics programs at the elementary (Schoen 1976[a]) and the secondary levels (Schoen 1976[b]) raised some doubts. In the year since these reviews were written, a more complete and updated survey of journals, doctoral dissertations, and the ERIC files has disclosed over one hundred studies, mostly unpublished, dealing with SPI mathematics programs. This essay will attempt to provide a fairly complete picture of the findings with some implications for future directions in individualized mathematics instruction.

Does SPI Improve Mathematics Achievement?

This is a complicated question for several reasons. First, since there are many SPI programs that vary to some extent in potentially important ways, there is no one answer. Results with one program will not necessarily be generalizable with those of other programs or future efforts in the direction of self-pacing. In this section different SPI programs are separated as much as possible to simplify interpretation. Second, many of the studies involved

SPI programs in their first year of operation. Obviously, difficulties peculiar to a new SPI program may have affected the results. To isolate this effect of the newness of the SPI program, longitudinal studies and those involving established SPI programs are separated in another section, "What Are the Long-Range Effects of SPI?" Third, mathematics achievement is composed of several factors. These are treated separately, and results with norm-referenced, standardized achievement tests and criterion-referenced measures are separated as well. Fourth, the effectiveness of SPI, as measured by student achievement, varied with the grade level of the students. To isolate this effect, separate analyses in five grade-level categories are made.

The following are characteristics common to nearly all the studies summarized here (some exceptions will be noted in the discussion of the individual studies):

1. Only studies in which the comparison groups were "equivalent" before the treatment are considered in this analysis. The methods of achieving equivalence were (a) randomly assigning students or classes to treatments; (b) matching one treatment group to another on several variables, such as sex, IQ, or previous mathematics achievement; or (c) statistically equating the groups by analysis of covariance or by using the difference between pretest and posttest scores as criterion measures.

2. The length of the studies was typically either one semester or one academic year.

3. The students comprising the samples were predominantly white, middle class students normally distributed on school variables as well as socioeconomic variables. Several exceptions to this are given separate consideration in a later section, "What Types of Students Are Successful in SPI?" The sample sizes ranged from sixty-two to over one thousand, with a median of about three hundred fifty.

4. The criterion measures (dependent variables) were typically standardized arithmetic achievement tests and their subtests. The specific test used in each study will not be reported here. The Iowa Test of Basic Skills, Stanford Achievement Tests, SRA Mathematics Battery, Sequential Test of Educational Progress, and Comprehensive Test of Basic Skills were some of the tests used. Analysis of variance and t tests were the most common statistical techniques used to test differences in group means. Unless otherwise stated, "statistically significant" means differences beyond the .05 probability level.

In most studies, the SPI treatment was compared to traditional instruction (TI). The researchers did not always describe this method in great detail. With some noted exceptions, the traditional method was teacher centered and teacher paced with common tests given at the same time to all twenty to thirty-five students within a self-contained classroom. Again, the traditional approach differed from one study to another, but these were common characteristics that marked the contrast to the SPI approach.

Results by SPI program

Kozak (1974) gathered information on thirty-six different SPI programs that had gained more than local use. Some of the most widely used SPI programs were Individually Prescribed Instruction (IPI), Program for Learning in Accordance with Needs (PLAN), and Individually Guided Education (IGE). Several evaluations of each of these programs were found, but in the majority of studies the SPI program was developed by the researcher or local school personnel. Table 1 lists the number of favorable, unfavorable, and "no significant difference" findings by SPI program and grade level.

TABLE 1

MATHEMATICS ACHIEVEMENT* IN SPI VERSUS TI CLASSES

NUMBER OF ANALYSES REPORTED BY OUTCOME CATEGORY

Grade Level	IPI			PLAN			IGE			Local			Total		
	Fav.	Opp.	NS	Fav.	Opp.	NS	Fav.	Opp.	NS	Fav.	Opp.	NS	Fav.	Opp.	NS
K–3	2	1	4	1	0	1	1	0	1	5	2	4	9	3	10
4–6	1	6	13	0	3	2	1	0	2	5	4	14	7	13	31
7–9	0	0	4	0	0	0	0	0	0	5	9	21	5	9	25
10–12	0	0	0	0	0	0	0	0	0	1	3	6	1	3	6
Coll	0	0	0	0	0	0	0	0	0	1	0	4	1	0	4
Total	3	7	21	1	3	3	2	0	3	17	18	49	23	28	76

* No distinction is made among total achievement and the various achievement subtests.

Overall, the evidence appears to favor TI, with over half the analyses resulting in no significant difference. The totals also indicate that SPI has been most effective with children in kindergarten through grade 3 and least effective in the intermediate and junior high school grades. The criteria were mathematics achievement tests and their subtests.

Some controversy has arisen concerning the appropriateness of using standardized achievement tests to measure student outcomes in SPI programs. No attempt will be made here to enter into that controversy. However, several researchers used criterion-referenced tests designed to measure the objectives of their SPI program. The results fit nearly the same pattern that is seen in table 1. Four analyses favored SPI (at the first-grade, third-grade, fourth-grade, and college levels), five favored TI (fourth-, seventh-, eighth-, and two eleventh-grade levels), and seven resulted in no significant difference.

Results by achievement subtest

Standardized mathematics achievement tests consist of subtests designed to measure separate factors of mathematics achievement: computation, concepts, applications, reasoning, and problem solving. No one standardized test contained all these subtests, but most contained three of them. Although a good deal of overlap exists in the applications, reasoning, and problem-solving subtests, the outcomes on the five subtests are listed separately in table 2. Norm-referenced, standardized arithmetic tests were not used in any studies with children beyond grade 9.

TABLE 2
ACHIEVEMENT SUBTEST RESULTS
NUMBER OF ANALYSES BY OUTCOME CATEGORY

Subtest	K–3			4–6			7–9			Total		
	SPI	TI	NS	SPI	TI	NS	SPI	TI	NS	SPI	TI	NS
Computation	1	1	1	1	8	2	0	2	4	2	11	7
Concepts	1	0	2	2	1	7	0	2	3	3	3	12
Applications	0	0	0	0	1	6	1	1	1	1	2	7
Reasoning	0	0	1	0	1	0	0	1	0	0	2	1
Problem Solving	1	0	0	0	0	1	0	1	1	1	1	2
Totals	3	1	4	3	11	16	1	7	9	7	19	29

The outcomes on the computation subtests, especially in the intermediate and junior high school grades, show a very clear TI superiority. In fact, the overall trend of TI superiority found in table 1 may be attributed to this weakness of SPI in developing computational skills. The combined results on all subtests except computation were five favoring SPI and eight favoring TI; twenty-two did not show significant differences. This may indicate that very little is taught in elementary school mathematics in SPI or TI programs that is tested by these subtests. Or perhaps the two instructional approaches were about equally good (or bad) at developing the necessary concepts and problem-solving skills.

Does SPI Improve Affective Student Outcomes?

It has often been argued that SPI has a more positive effect on a student's attitude, self-concept, and general feeling of well-being than TI has. Many of the comparative studies included in the compilation of achievement results in the previous section also provided a comparison based on one or more affective criteria. The findings were again mixed and depended on the grade level of the subjects. In the interest of saving space, no attempt will be made here to describe the individual affective categories and testing instruments. Rather, the overall trends will be shown. These findings are summarized in table 3. Included in the "Other" category are adjustment to school, creativity, resourcefulness, study habits, spontaneous and adaptive flexibility, and anxiety about school.

TABLE 3
AFFECTIVE CRITERIA
NUMBER OF ANALYSES REPORTED BY OUTCOME CATEGORY

Criterion	K–3			4–6			7–9			10–12			College			Total		
	SPI	TI	NS	SPI	TI	NS	SPI	TI	NS	SPI	TI	NS	SPI	TI	NS	SPI	TI	NS
Attitude toward School	4	0	2	0	1	5	0	0	2	0	1	0	0	0	0	4	2	9
Attitude toward Mathematics	0	1	1	0	0	1	1	0	6	0	0	5	0	0	3	1	1	16
Self-Concept	1	0	1	1	0	6	0	0	2	0	0	0	0	0	0	2	0	9
Other	3	0	0	0	1	5	0	0	0	0	0	0	1	0	1	4	1	6
Total	8	1	4	1	2	17	1	0	10	0	1	5	1	0	4	11	4	40

The grand totals show SPI to be superior to TI in eleven of fifty-five analyses, whereas the reverse was true in only four cases. However, eight of the analyses favoring SPI were at the primary-grade levels. Above the third-grade level, nearly all analyses resulted in no significant difference between SPI and TI. This may reflect an inability to measure affective variables rather than the lack of real differences.

What Are the Long-Range Effects of SPI?

An educational innovation as massive as the adoption of an SPI mathematics program is a difficult undertaking. Time is required to smooth over the rough spots in the materials and management system, to orient the teachers to their new role, to hire and train qualified teacher aides, and to give the students an opportunity to become comfortable with the less restricted classroom atmosphere and the new approach to learning and evaluation. Premature evaluation for purposes of accepting or rejecting a particular SPI mathematics program may result in a costly mistake.

Most of the comparative studies tabulated in the previous two sections were evaluations of SPI programs in the first year of operation, with comparisons being made over a period of one semester to a year. However, there were enough exceptions to this pattern to make separate analyses of longitudinal studies and studies involving SPI programs that had been in operation for one or more years prior to the reported evaluation. The results of evaluations of SPI programs that were in at least their second year of operation are categorized in table 4.

TABLE 4
SPI PROGRAMS IN AT LEAST SECOND YEAR
NUMBER OF ANALYSES BY OUTCOME CATEGORY

Criterion	K–3			4–6			7–9			10–12			College			Total		
	SPI	TI	NS	SPI	TI	NS	SPI	TI	NS	SPI	TI	NS	SPI	TI	NS	SPI	TI	NS
Mathematics Achievement*	0	2	3	1	3	9	1	3	3	0	0	2	0	0	0	2	8	17
Affective Criteria	3	1	1	0	1	6	1	0	3	0	1	0	0	0	0	4	3	10

* No distinction is made among achievement subtests and total achievement.

Several interesting trends can be seen by comparing table 4 with tables 1 and 3. First, seventeen of the twenty-seven comparisons on mathematics achievement measures resulted in no significant difference. Second, the differing pattern of results by grade levels is not as clear as it is when all studies are considered. In particular, the superiority of SPI over TI in kindergarten through grade 3 has disappeared entirely. The trends in achievement still favor TI in the intermediate and junior high school grades, but they are less pronounced than in table 1, where all studies were considered. Third, there is little evidence to support the claims of SPI proponents in the affective domain. The clear-cut superiority of SPI in the primary grades evident in table 3 is no longer so clear, and there is little overall difference between SPI and TI with older students.

Four longitudinal studies, two over a three-year period (Helms 1974; Taylor and Fleming 1972) and two over a five-year period (Morris 1969; Cross 1974), also show a trend toward the same or lesser mathematics achievement in SPI. Only one longitudinal comparison of affective outcomes was made. Morris (1969) found that high school students' attitude toward school became more negative after four years of SPI. Explanations other than the ineffectiveness of SPI may be given for some of the negative findings, but there is still a paucity of evidence showing that students improved in either the cognitive or affective domains when they were enrolled in SPI programs.

What Types of Students Are Successful in SPI?

In addition to the comparative studies, much recent research has dealt with identifying student traits that will predict success in SPI and with the interaction of student characteristics with instructional treatment (SPI versus TI). Several fairly consistent findings have emerged:

1. Children in the primary grades are more likely to react positively to SPI than older children. Their arithmetic achievement rate often improves initially, but that improvement is not maintained.

2. SPI has a differential effect on students of different levels of IQ or mathematical ability. In particular, SPI tends to decrease the mathematical achievement rate of low-ability students while sometimes improving their attitude toward school. A low reading level, an inability to keep accurate records, a reluctance to seek help, a need for the security of teacher guidance, and a lack of self-motivation are the most likely explanations for the poor showing of low-ability students in SPI. High-ability students tend to do about as well with SPI as with TI.

3. When one uses attitude or achievement criteria, there is no consistent evidence that there is any sex by treatment (SPI versus TI) interaction at any grade levels.

4. Whether or not the reading ability of the students is related to success in SPI appears to depend on the nature of the SPI program. Some programs require more reading than others.

5. SPI used in situations with exceptionally low teacher/pupil ratios meets with fairly consistent success when the students have special learning problems (e.g., educable mentally retarded or learning disabled students). Perhaps the careful diagnosis of individual weaknesses and the relatively low level of the learning objectives are compatible with their needs. For whatever reasons, the profile of results with these students looks better than it does with the school population in general.

6. The student characteristics that have been the best predictors of success in SPI are a high ability level and high self-motivation. When given the option, these students have also been more likely than others to choose SPI over TI.

How Is Student Behavior Affected by SPI?

In a broad sense, achievement and affective measures are indicators of student behavior. However, in this section the study skills, study habits, and strategies for student learning in SPI classrooms will be examined.

In a five-year study of a high school SPI program, Morris (1969) found that on the one hand the high school dropout rate decreased significantly. On the other hand, study habits and library skills of students in SPI were poorer than they had been with TI. The latter finding may be a result of that particular SPI program, which does not require library work. Also it is not surprising that students' study habits would be different, if not poorer, with SPI. The mastery-learning approach with carefully specified objectives and frequent testing requires a change in study habits. The lower dropout rate is a high priority item in many schools, but one should be careful about attributing it to SPI in the Morris study. The study involved only one high school, and no description of any other changes in school policies that may have contributed to the decrease in the dropout rate was given.

In many SPI programs students are responsible for scoring tests and worksheets designed to help them determine if they need remediation or if they should proceed to the unit's posttest. Students are trusted to use the key to check their work, to attempt to rework any incorrect problems, and then to see the teacher for help if it is needed. This is an important component of the program, since it provides the student with feedback and diagnostic data. Oles (1973) found that only about 12 percent of the IPI fifth graders in his sample followed the appropriate procedure when confronted with incorrect answers. More often, incorrect answers either were not marked wrong and ignored or were marked wrong but not reworked. Oles found that students enjoyed self-scoring; they admitted to cheating at times but did not think it was wrong.

Student practices of this sort may help explain the seemingly low quality of mathematics learning with SPI reported by several researchers. For example, Erlwanger (1975) conducted case studies of two sixth-grade boys, Benny and Mat, who were near the top of their IPI arithmetic classes in both rate and quality of their work. Both boys had been in the program since second grade. Both were found to have very basic misconceptions in mathematics and to view mathematics as a game in which they had to guess the answer in the IPI key. Both boys felt that the right answer depended on what was in the key, not on the logic of the mathematics problem. When the interviewer pointed out misconceptions, the boys were very reluctant to attempt to relearn any previous content since in their view, they had finished that material. An excerpt from an interview with Benny (*B.*), illustrates this viewpoint (Erlwanger 1975, p. 132):

> *E.*　It [finding answers] seems to be like a game.
>
> *B.*　[Emotionally] Yes! It's like a wild goose chase.
>
> *E.*　So you're chasing answers the teacher wants?
>
> *B.*　Ya, ya.
>
> *E.*　Which answers would you put down?
>
> *B.*　[Shouting] Any! As long as I knew it could be the right answer. You see, I am used to check my own work, and I am used to the key. So I put down ½ because I don't want to get it wrong.
>
> *E.*　Mm. . .
>
> *B.*　Because if I put ¼ and ¼, they'll mark it wrong. But it would be right. You agree with me there O.K. If I put 2/4, you agree there. If I put ½, you agree there too. They're all right! . . . They mark it

wrong because they just go by the key. They don't go by if the answer is true or not. They go by the key. It's like if I have 2/4. They wanted to know that it was, and I write down one whole number, and the key said a whole number, it would be right; no matter if it was wrong.

Mat, a second highly successful IPI sixth grader, had to select the correct answer to the following exercise in his booklet:

$$32/4 \text{ is equal to } 8,\ 2\ 1/5,\ 32 \times 4,\ 32 \div 4$$

He explained his reasons for selecting 32×4 and $32 \div 4$ as follows (Erlwanger 1975, p. 139): "Well, because 32 over 4. . . . It doesn't make sense for it to be an 8, and this is what the first one is. So I didn't circle it. And 2 1/5 . . . well, to me what I did, if the problem doesn't make sense, I don't circle it . . . because 32 over 4 doesn't have anything to do with 2 1/5. And 32 $\times$ 4, the 32 is on the top and it's right here, and then it's time 4 down here, and a 4 is on the bottom there. And 32 divided by 4 is the same thing."

It is not uncommon for sixth graders to have such misconceptions and rule-oriented views of mathematics, but Mat and Benny were highly successful students who held very rigidly to their viewpoints even after several individual remedial sessions. If these difficulties are due to the IPI program, what aspects of it need to be improved? Perhaps minor improvements are needed, or perhaps IPI (or SPI in general) is an inadequate instructional approach. Research dealing with components of the SPI program may provide some help in finding an answer to these questions. The findings to date are considered in the next sections.

What Is the Teacher's Role in SPI?

Virtually every researcher agreed that without some form of assistance teachers have more work with SPI than with TI. For example, Nix (1970) found that SPI teachers spent 1.8 times as many total hours in daily preparations, preparing examinations, checking homework, and grading examinations as TI teachers. To counteract the problem of overworked teachers, many commercially available SPI programs recommend one or two teacher aides for every twenty-five students. It appears that a prerequisite for an SPI program to enjoy long-term success is at least one of the following: (*a*) a teacher aide (or aides) with each group of students, (*b*) a greatly lowered student/teacher ratio, or (*c*) a computer-managed system of record keeping.

What do teachers do in the SPI classroom? In the ideal program they are available to help individual students as they are needed, to discuss the students' work with them individually, to diagnose individual learning problems, to prescribe each student's course of study based on diagnosis and student input—in short, they are managers and facilitators of learning. They are not disciplinarians and dispensers of information, as the teacher in TI is often characterized.

How realistic is this description of the SPI teacher? Neujahr (1971) videotaped four sixth-grade SPI classrooms over a six-week period. He found that five-sixths of the interaction between the teacher and student involved procedural matters, compared to one-third in the typical TI classroom. In addition, the interaction that was of a substantive educational nature was always very specific to the topic at hand. This seems to be more restrictive than the TI classroom, in which typically one-eighth of the interaction is off the specific topic at hand but within the subject. Similarly, Sutton (1967) found that SPI seventh graders raised twice as many questions about procedural matters as about mathematics.

If these researchers have provided an accurate picture of the teacher's role in SPI, this is certainly another serious problem that may help explain some of the poor student outcomes. The simplest solutions (but expensive ones) seem to be those mentioned earlier: teacher aides, lower student/teacher ratio, and a computer-managed record-keeping system. The physical arrangement of the classroom, the availability of materials, and the efficiency of the management system can all be factors affecting the number of procedural matters with which the teacher must deal during class time. Organizational ability and efficiency become very important characteristics of a good SPI teacher.

What Is the Best Approach to Diagnosis and Prescription in SPI?

Most SPI programs depend on a diagnosis of the individual students' entering achievement level and learning problems in order to place them in a mode of instruction that is judged to be compatible with their personality and learning style. The mechanism for doing this is curriculum-embedded diagnostic tests interpreted by the teacher following the curriculum developers' guidelines. Prescription procedures range from very specific matrices of objectives keyed to the test items to the teacher's judgment of the student's personality or learning style. Researchers have found that this component of SPI has not been effective for several reasons:

1. Most teachers tended to prescribe almost all the instruction in the next unit to almost all the students. There seemed to be a fear that otherwise something important would be missed (Graeber 1974).

2. One study compared four prescription strategies in seven primary IPI mathematics classrooms. Students whose teachers used the full array of prescription options recommended by the IPI developers scored lowest

on the curriculum-embedded mastery and retention tests. Students who were given no prescription at all scored the best of the four treatment groups (Holste 1972).

3. When left to choose their own assignment levels without guidance from the teacher, seventh-grade students tended to choose one level and continue in it all year long. Only high-ability students were able to choose their correct level (Snyder 1967).

4. Placing students into instructional modes according to their personality characteristics has not been shown to have a beneficial effect on mathematics achievement in SPI. For example, Stiglmeier (1973) identified eighth-graders who had been consistently judged by several teachers to be either highly self-reliant or highly dependent. These students were placed in self-supportive and teacher-supportive modes of instruction, respectively, within an SPI mathematics program. Another group of eighth-graders were similarly judged highly self-reliant or highly dependent but were placed randomly into an instructional mode. No significant difference in the mean mathematics achievement after one year of instruction was found between the matched group and the random group. In fact, all students achieved more in teacher-supportive modes regardless of their characteristics.

5. It seems that teachers would differ in their ability to make good prescriptions. However, DeRenzis (1971) found little difference between "control oriented" and "freedom oriented" teachers' ability to write good prescriptions, nor did this ability improve with years of SPI experience.

The research points to diagnosis and prescription as an area of weakness in SPI programs. Perhaps the problem is one of teacher training. Most commercial SPI programs do recommend in-service teacher training, but many researchers point to a need for improvement in this area. Even with more training the prescription aspect of SPI is not likely to improve much, at least not immediately, since little is known about the measurement of learning styles and how best to match them to modes of instruction. More research with practical implications is needed before anyone will be able to design effective strategies in this area.

However, SPI developers should seriously consider the findings and recommendations of Moncrief (1973) if they wish to improve the reliability of their prescription strategies and instructional sequence. He found that only 15 percent of the units in the Individualized Mathematics System (IMS) had valid performance criteria. This was evidenced by the fact that students who failed to meet the criterion on a unit's posttest were often able to succeed in the next units and, in fact, later raised their scores on a retest of the unit on which the failure had occurred. Thus, a good deal of the time and effort of both the teacher and the student were being wasted in a unit when the optimal strategy was to allow the student to proceed to the next unit. Moncrief recommended that SPI developers set performance criteria empirically and not by the usual a priori guesswork.

How Useful Are Supplementary Materials and Teaching Aids in SPI?

There is consistent evidence that the use of a wide variety of supplementary materials and manipulative devices can improve mathematics achievement in SPI programs. Unfortunately, there is also evidence that these materials often have not been available and have been rarely used when they were available. Herein may lie another reason for the poor student outcomes in SPI.

There were probably several reasons why a more extensive use of materials was not made in SPI. For one thing, many of the commercial programs have a small variety of materials, usually only printed matter and audiotapes (Edling 1970). Other factors contributing to the disuse of materials lay with the teachers and students themselves. Neujahr (1971) found in his analysis of the SPI classroom environment that only one-seventh of the teacher-initiated moves focused on the use of materials in the program, whereas three-sevenths focused on the work the pupil should be doing. Perhaps teachers have tended to discourage the use of multimedia and supplementary materials, viewing them as peripheral to the student's real objectives or as more trouble than they were worth. Students, too, may have discovered that they could pass the required tests more efficiently without using the manipulative devices and teaching aides.

How Much Does SPI Cost?

In his analysis of thirty-six SPI programs, Kozak (1974) reported that the cost of noncomputerized programs ranged from $15 to $115 a year for each student, excluding the cost of teacher training and building renovation. Computerized programs ranged from $48 to $144 a year for each student. The cost of materials alone averaged $3.20 a year for each student when projected over a five-year period. This is compared to $0.80 a student a year for a textbook, just one-fourth of the cost of SPI materials. The average costs reported by Kozak were collected from the SPI program developers and varied somewhat from those reported by individual researchers.

Several researchers reported the costs of IPI programs on the basis of price per student per year. Principals of nine Illinois schools using IPI reported the cost of materials ranging from $6.50 to $9.00 (Karmos 1975). This cost did not include the salaries of one or two teacher aides in each classroom in eight of the schools. In another IPI school, the cost of materials decreased with each year of operation but was still quite high after four years (Meade and Griffin 1969). The figures for the first four years of operation were $42, $18, $16, and $12 respectively. However, Taylor and Fleming (1972) found that the extra total cost of IPI over TI increased each year of the first three years of operation. Reported excess costs were $121, $162, and $210 respectively. These included the salaries of two teacher aides in each classroom. Harper (1973) found that the total IPI materials costs were five times the cost of TI in his comparative study. Finally, in an article describing the advantages of the IPI system, Edmunds (1971) wrote, "At present the IPI maths program costs about $10 a year per child compared with about $1.50 for conventional maths programs. Costs, of course, will drop as the programs go into commercial production, but they will still remain higher than traditional programs because of the variety of materials involved" (p. 13).

Costs of other SPI programs were also documented. In a report of an early formative evaluation study of IMS, $3.20 was projected as an average cost for materials over five years (Frary 1971). One of the most prohibitive costs reported was $425 more than TI for a computer-assisted program using a wide variety of materials and many teacher aides (Broussard 1971).

Another factor in figuring cost is the renovation of buildings. This cost will vary depending on the extent of the renovation. One researcher reported that a junior high school spent $9000 to remove all the walls to adopt the Denver Continuous Progress Mathematics Program (Colvin 1973).

There is no doubt that SPI is more expensive than TI. The cost of SPI materials alone will be about four or five times those for TI. To this must also be added the cost of extra personnel as well as building renovations. Whether or not the cost is justified must be a decision based on the priorities of individual school systems. LaPlaca (1974) attempted to show the cost effectiveness of an eighth-grade SPI mathematics program based on mathematics achievement. Although the SPI group mean was significantly higher than that of the TI group, the cost of SPI was $0.54 a unit gained in pretest to posttest achievement compared to $0.36 a unit for TI. This finding and the typically poor student outcomes on achievement measures make it highly unlikely that SPI, at its present stage of development, can be shown to be cost effective when based on the achievement variable alone.

Summary and Implications

The results of many studies have been summarized. To condense them even further is to oversimplify the findings and their implications. The points in this section, even more than in the preceding ones, should be viewed with the following limitations in mind:

a. The SPI programs evaluated, as well as the TI approaches, varied in many important ways. The tabulated outcomes provide a broad profile of findings to date. They may or may not have any application to a particular program in a particular setting.

b. Although the findings have been condensed, every attempt has been made to keep the interpretations as objective as possible. The best and most complete picture of an individual researcher's findings can be found by using the list of references.

c. These findings are specific to SPI mathematics programs and mathematics achievement. The results may be entirely different with other subject matter.

The purpose of this essay was to provide a summary of recent research concerning SPI mathematics programs and to consider some implications of the findings. Several questions of importance to mathematics teachers and school administrators were posed, and the findings relative to each question were discussed. A brief summary follows.

Does SPI improve mathematics achievement? Using mathematics achievement as the criterion, researchers found that results favored TI more often than SPI, although many analyses resulted in "no significant difference." No one SPI program was able to show a reversal of that trend, and locally developed programs were about as effective as those sold commercially. This pattern of findings also holds true when tests keyed to the objectives of the SPI program were used instead of standardized achievement measures. SPI was particularly ineffective in developing computational skills at the intermediate and junior high school levels.

Does SPI improve affective student outcomes? On affective criteria, forty of fifty-five analyses resulted in no significant difference between the means of the TI and SPI groups. SPI groups were consistently superior to TI groups only at the primary-grade levels.

What are the long-range effects of SPI? If studies over at least a three-year period and other evaluations of the SPI programs in at least their second year of operation are considered, a pattern similar to that described under the first question emerges in achievement at all levels. However, the pattern of results does not vary by grade level, and there appears to be little difference between SPI and TI when results are based on affective criteria.

What types of students are sucessful in SPI? Entering ability level and self-motivation have been the best predictors of a student's success in SPI. Generally, high-ability students achieved equally as well in SPI as in TI, but most low-ability students were unable to function in SPI. Variations of SPI have been used effectively with students who have special learning problems.

How is student behavior affected by SPI? In one five-year study, SPI high school students showed a decrease on measures of study habits and library skills, but the school dropout rate also decreased. In another study, students were found to cheat regularly when scoring their curriculum-embedded tests. In case studies, two high-achieving sixth-grade students who had been in IPI for four years were found to have very basic misconceptions in mathematics and a rigid, "get the keyed answer" view of mathematics learning.

What is the teacher's role in SPI? Teachers in SPI have tended to spend more time dealing with procedural matters and to restrict educational discussion more specifically to the topic at hand than teachers in a typical TI classroom. Because of excessive amounts of management and record-keeping demands, SPI programs need one of the following: (*a*) teacher aide(s) with each group of students, (*b*) a greatly lowered student/teacher ratio, or (*c*) a computer-managed record-keeping system.

What is the best approach to diagnosis and prescription in SPI? Several approaches have been tested, but none was shown to be superior to any other, nor was any shown to be superior to using no diagnosis and prescription at all. In fact, the diagnosis-prescription aspect of SPI has not been shown to be effective. One difficulty with prescription writing may lie in the fact that the criteria for the units have been invalid.

How useful are supplementary materials and teaching aids in SPI? There is consistent evidence that the use of various media and supplementary teaching materials increased the effectiveness of SPI. There is also consistent evidence that media and materials in a typical SPI program have been restricted to printed and audio materials. In addition, the various media and supplementary materials often have not been used even when available.

How much does SPI cost? SPI programs certainly cost more than TI. Sources of the extra expenses are materials, teacher aides, teacher training, and sometimes computer service and building renovation. Projected over a five-year period, costs for materials alone for SPI were about four or five times that of TI.

To draw all these findings together to make specific recommendations is very difficult, but several points emerge. First, SPI is one approach to meeting the needs of individual students. Its advocates and practitioners have worked hard and enthusiastically in their attempt to make it a workable and effective approach. However, they will be the first to admit that

SPI programs are expensive and that a good deal of time and effort is required from school personnel to adopt and operate them. These facts are also consistent with the research findings. Surely some extra educational benefits should result from the expense and effort.

Second, thus far the research has failed to demonstrate consistently a superiority of SPI over TI on any student-outcome variables. On achievement variables the findings favor TI more often, and on affective variables there appears to be little difference. It may be argued that the right variables have not been measured, the particular SPI programs or their management systems were ineffective, or that we do not know how to measure the variables that would show the superiority of SPI. Each of these arguments has a degree of validity for individual studies, but sooner or later school administrators, mathematics teachers, and the American taxpayers will and should demand a demonstrated return for the extra investment of money and effort.

A third important point is that SPI appeared to be as effective as TI with the self-motivated, high-ability students and those with special learning problems. A TI mathematics classroom teacher or an entire school could achieve a degree of individualization by providing SPI for those groups of students and by providing more teacher-supportive instruction for the others.

Fourth, some practices in SPI appear to have promise for improving student outcomes. One is the availability and the use of multimedia options and supplementary teaching materials. The development of unit criteria based on empirical evidence is another. In general, practices that have provided more teacher guidance have increased student achievement also.

Fifth, several particularly weak areas of SPI have been identified. Some of these are diagnosis and prescription, unreliable student self-scoring, the development of computational skills in intermediate and junior high school grades, meeting the needs of low-ability students, and the quality of conceptual learning. Improvements are needed in these areas if future SPI programs are to be more effective than those evaluated thus far.

Sixth, there are many approaches to individualizing mathematics instruction that do not fit the SPI model; Willoughby (1976) lists a total of twenty. For the most part, the effectiveness of each of these approaches is an open question. More research is needed before the strengths and weaknesses of these various individualization schemes can be judged.

Finally, the findings that are summarized here should not be viewed as a reason to forget the individual student. Meeting individual needs should be of utmost importance to mathematics educators. However, the SPI approach to individualization has many weaknesses that detract from its overall effectiveness. By isolating specific areas of strength and weakness, this essay is meant to provide a basis for new and more effective directions for individualizing mathematics instruction. Albert Einstein's remark concerning Herbert Spencer's idea of tragedy—"a deduction killed by a fact"—applies here: "Every theory is killed sooner or later in that way. But if the theory has good in it, that good is embodied and continued in the next theory" (R. W. Clark, *Einstein, the Life and Times* [New York: World Publishing Co., 1971], p. 624).

CITED REFERENCES

Broussard, Vernon. *The Effect of an Individualized Instructional Approach on the Academic Achievement in Mathematics of Inner-City Children.* (Doctoral dissertation, Michigan State University, 1971.) Ann Arbor, Mich.: University Microfilms (no. 71-31,166), 1971.

Colvin, Dan. *Evaluation of CPMP (Grades 6–12, Continuous Progress Mathematics Program) 1972–1973.* Final report of an ESEA Title III project, 1973. (ERIC no. ED 085 400)

Cross, Myron E. *An Evaluation of the Individualized Instructional Program of the Natomas Union School Districts.* (Doctoral dissertation, Brigham Young University, 1974.) Ann Arbor, Mich.: University Microfilms (no. 74-18,250), 1974.

DeRenzis, Joseph J. *An Investigation into the Attitude Patterns and Their Relationship to Prescription Writing Procedures of Teachers Using the IPI Instructional System in Elementary Mathematics.* (Doctoral dissertation, Temple University, 1970.) Ann Arbor, Mich.: University Microfilms (no. 71-10,811), 1971.

Edling, Jack V. "Individualized Instruction: The Way It Is in 1970." *Audiovisual Instruction* 15 (1970): 13–16.

Edmunds, J. "Individually Prescribed Instruction." *Orbit* 2 (1971): 10–13.

Erlwanger, Stanley. "The Observation Interview Method and Some Case Studies." In *Proceedings of the Conference on the Future of Mathematical Education.* Tallahassee, Fla.: Florida State University, 1975.

Frary, Robert B. *Formative Evaluation of the Individualized Mathematics System (IMS).* Durham, N.C.: National Laboratory of Higher Education, 1971.

Graeber, Anna O. "Diagnostic-Test-based Prescriptions in Individually Prescribed Instruction in Mathematics." (Doctoral dissertation, Columbia University.) *Dissertation Abstracts* 35A (1974): 2646.

Grittner, Frank M. "Individualized Instruction: New Myths and Old Realities." *Wisconsin Journal of Public Instruction* (Winter 1971): 49–59.

Harper, Kenneth J. *A Comparison of Three Elementary Mathematics Programs: A Model for Curriculum Evaluation.* (Doctoral dissertation, Wayne State University, 1972.) Ann Arbor, Mich.: University Microfilms (no. 73-12,526), 1973.

Helms, David C., Jr. "Use of a Formative Evaluation Technique in Determining the Differential-Achievement Effects of the IPI Mathematics, 1967–1970." (Doctoral dissertation, Temple University.) *Dissertation Abstracts* 34A (1974): 4698–99.

Holste, Donald E. *The Effect of Different Prescriptions Used by Teachers in an IPI Mathematics Program.* (Doctoral dissertation, University of Illinois, 1972.) Ann Arbor, Mich.: University Microfilms (no. 72-19,849), 1972.

Karmos, Joseph S. *A Study of Four Factors Associated with the Installation of the Individually Prescribed Instruction in Mathematics Program in Ten Illinois Schools.* (Doctoral dissertation, Southern Illinois University, 1974.) Ann Arbor, Mich.: University Microfilms (no. 75-13,274), 1975.

Kozak, Michael R. *A Critical Analysis of Individualized Instruction since 1944.* (Doctoral dissertation, Texas A & M University, 1974.) Ann Arbor, Mich.: University Microfilms (no. 75-2912), 1974.

LaPlaca, Nicholas A. "A Cost-Effectiveness Analysis of Individual Learning Units in a Junior High School Basic Mathematics Program." (Doctoral dissertation, University of the Pacific.) *Dissertation Abstracts* 34A (1974): 3771.

Meade, William F., and Lawrence M. Griffin. *A Comparative Study of Student Achievement and Other Selected Student Characteristics in a Program of Individualized Instruction in Mathematics and in a Program of Traditional Instruction in Mathematics in Grades 1–6.* Final report of an ESEA project, 1969. (ERIC no. ED 037 362)

Moncrief, Michael H. *A Validation Study of Selected Decision Rules Used in the Management of Student Progress through an Individualized Mathematics System.* (Doctoral dissertation, Florida State University, 1972.) Ann Arbor, Mich.: University Microfilms (no. 73-4695), 1973.

Morris, Julian C. "A Descriptive Analysis and Evaluation of an Integrated Program of Individualized Instruction in Cedar City High School." (Doctoral dissertation, Brigham Young University.) *Dissertation Abstracts* 29A (1969): 2937.

Neujahr, James L. *An Analysis of Teacher-Pupil Interactions When Instruction Is Individualized.* (Doctoral dissertation, Columbia University, 1970.) Ann Arbor, Mich.: University Microfilms (no. 71-5593), 1971.

Nix, George C. *An Experimental Study of Individualized Instruction in General Mathematics.* (Doctoral dissertation, Auburn University, 1969.) Ann Arbor, Mich.: University Microfilms (no. 70-1933), 1970.

Oles, Henry J. "Assessment of Student Self-Evaluation Skills." *Programmed Learning and Educational Technology* 10 (1973): 360–63.

Schoen, Harold L. (a). "Self-paced Mathematics Instruction: How Effective Has It Been?" *Arithmetic Teacher* 23 (February 1976): 90–96.

———— (b). "Self-paced Mathematics Instruction: How Effective Has It Been in Secondary and Postsecondary Schools?" *Mathematics Teacher* 69 (May 1976): 352–57.

Snyder, Henry D. *A Comparative Study of Two Self-Selection-pacing Approaches to Individualizing Instruction in Junior High School Mathematics.* (Doctoral dissertation, University of Michigan, 1966.) Ann Arbor, Mich.: University Microfilms (no. 67-8346), 1967.

Stiglmeier, Lois M. *Teachers' Judgments of Pupils' Dependence/Self-Reliance Characteristics Mode of Instruction and Their Relationship to Achievement.* (Doctoral dissertation, State University of New York at Albany, 1972.) Ann Arbor, Mich.: University Microfilms (no. 73-19,707), 1973.

Sutton, Joseph T. *Individualizing Junior High School Mathematics Instruction: An Experimental Study.* Final report of project no. 1365. Contract no. OE2-10-083, 1967. (ERIC no. ED 016 609)

Taylor, Derek B., and Margaret Fleming. *Individually Prescribed Instruction Program (Mathematics).* Disadvantaged Pupil Program Fund no. 97-19, 1971–72 evaluation, 1972. (ERIC no. ED 077 700)

Willoughby, Stephen S. "Individualization." *Mathematics Teacher* 69 (May 1976): 338–44.

UNCITED REFERENCES

Abate, Edwin C. *An Evaluation of an Individualized Educational System in an Elementary School.* (Doctoral dissertation, Columbia University, 1972.) Ann Arbor, Mich.: University Microfilms (no. 73-2575), 1973.

Acquaviva, Vincent M. "Provisions for Individual Differences in Mathematics Instruction in the Public High Schools of New Jersey." (Doctoral dissertation, Rutgers University.) *Dissertation Abstracts* 34A (1974): 4074.

Amendola, Anthony A. *Changes in Attitude and Achievement Effected by a Continuous Progress Education Program at the Elementary School Level.* (Doctoral dissertation, Arizona State University, 1973.) Ann Arbor, Mich.: University Microfilms (no. 73-5304), 1973.

Arrants, Glen C. *An Individually Prescribed Instructional Program Using Behavioral Objectives with Primary Educable Mentally Retarded Children.* (Doctoral dissertation, Duke University, 1972.) Ann Arbor, Mich.: University Microfilms (no. 73-6546), 1973.

Bartel, Elaine V. *A Study of the Feasibility of an Individualized Instructional Program in Elementary School Mathematics.* (Doctoral dissertation, University of Wisconsin, 1965.) Ann Arbor, Mich.: University Microfilms (no. 65-14,846), 1966.

Bazik, A. Matthew. *Evaluation of a Plan for Individualizing Instruction through Informing Students of Behavioral Objectives in a Mathematics Course for Prospective Elementary School Teachers at Elmhurst College.* (Doctoral dissertation, Northwestern University, 1972.) Ann Arbor, Mich.: University Microfilms (no. 73-10,181), 1973.

Beul, Bobbie T. "An Evaluative Study of Teaching Seventh-Grade Mathematics Incorporating Team Teaching, Individualized Instruction and Team Supervision Utilizing the Strategy of Learning for Mastery." (Doctoral dissertation, Saint Louis University.) *Dissertation Abstracts* 34A (1974): 4685.

Bledsoe, Joseph C., Jerry D. Purser, and Nevin R. Frantz, Jr. "Effects of Manipulative Activities on Arithmetic Achievement and Retention." *Psychological Reports* 35 (1974): 247–52.

Bowen, Robert L. "An Evaluative Study of an Individualized Math Team Program." (Doctoral dissertation, University of Southern California.) *Dissertation Abstracts* 34A (1974): 6349.

Bradford, Equilla F. *A Comparison of Two Methods of Teaching in the Elementary School as Related to Achievement in Reading, Mathematics, and Self-Concept of Children* (Doctoral dissertation, Michigan State University, 1972.) Ann Arbor, Mich.: University Microfilms (no. 73-5334), 1973.

Bronder, Cecilia C. *The Application of Diagnostic Teaching and a Mathematic [sic] Laboratory to a Middle School Individualized Unit on Fractions.* (Doctoral dissertation, University of Pittsburgh, 1973.) Ann Arbor, Mich.: University Microfilms (no. 73-24,091), 1973.

Brust, Joseph V. *The Relationship of Individualized Instruction in Learning Skills to Self-Esteem and Achievement.* (Doctoral dissertation, Columbia University, 1972.) Ann Arbor, Mich.: University Microfilms (no. 72-19,510), 1972.

Bull, Scott S. *A Comparison of the Achievement of Geometry Students Taught by Individualized Instruction and Traditional Instruction.* (Doctoral dissertation, Arizona State University, 1971.) Ann Arbor, Mich.: University Microfilms (no. 71-5976), 1971.

Burchyett, James A. *A Comparison of the Effects of Nongraded, Multi-Age, Team Teaching vs. the Modified Self-contained Classroom at the Elementary School Level.* (Doctoral dissertation, Michigan State University, 1972.) Ann Arbor, Mich.: University Microfilms (no. 73-12,686), 1973.

Chatterley, Louis J. *A Comparison of Selected Modes of Individualized Instruction in Mathematics for Effectiveness and Efficiency.* (Doctoral dissertation, The University of Texas at Austin, 1972.) Ann Arbor, Mich.: University Microfilms (no. 73-7532), 1973.

Clough, Roger A. *An Analysis of Student Achievement in Mathematics When Individually Prescribed Instruction (IPI) Is Compared to the Current Instructional Program.* (Doctoral dissertation, University of Nebraska, 1971.) Ann Arbor, Mich.: University Microfilms (no. 71-28,604), 1971.

Corbin, Harold G. "An Individualized Approach: An Evaluation of Cognitive and Affective Learning in Seventh and Eighth Grade Mathematics Classes." (Doctoral dissertation, University of Southern California.) *Dissertation Abstracts* 34A (1974): 6939.

Corn, J., and A. A. Behr. "A Comparison of Three Methods of Teaching Remedial Mathematics as Measured by Results in a Follow-up Course." *MATYC Journal* 9 (1975): 9–13.

Crandall, Larry D. "The Effects of Peer Tutors and Individual Skill Kits on Arithmetic Achievement and Attitude in Grade Seven." (Doctoral dissertation, University of Michigan.) *Dissertation Abstracts* 35A (1974): 94–95.

Crangle, Eva A. *An Evaluative Study of the Northwest Junior High School Individualized Mathematics Program.* (Doctoral dissertation, University of Utah, 1971.) Ann Arbor, Mich.: University Microfilms (no. 71-25,007), 1971.

Dahlke, Richard M. "Determining the Best Predictors of Success and of Time of Completion or Dropout in an Individualized Course in Arithmetic at a Community College." *Journal for Research in Mathematics Education* 5 (November 1974): 213–23.

————. "Studying the Individual in an Individualized Course in Arithmetic at a Comunity College: A Report on Four Case Studies." *Mathematics Teacher* 68 (March 1975): 181–88.

Deep, Donald. "The Effect of an Individually Prescribed Instruction Program in Arithmetic on Pupils at Different Ability Levels." (Doctoral dissertation, University of Pittsburgh.) *Dissertation Abstracts* 27A (1967): 2310.

Earnshaw, George L. *Open Education as a Humanistic Intervention Strategy.* (Doctoral dissertation, Syracuse University, 1972.) Ann Arbor, Mich.: University Microfilms (no. 73-19,801), 1973.

Emery, Harriett E. *Mathematics for Prospective Elementary Teachers in a Community College: A Comparison of Audio-Tutorial and Conventional Teaching Materials and Modes.* (Doctoral dissertation, Michigan State University, 1970.) Ann Arbor, Mich.: University Microfilms (no. 71-11,828), 1971.

Englert, Thomas J. *A Comparative Study of the Effects on Achievement and Changes in Attitude of Senior High School Students Enrolled in First Year Algebra under Two Different Teaching Approaches.* (Doctoral dissertation, Cornell University, 1972.) Ann Arbor, Mich.: University Microfilms (no. 72-23,657), 1972.

Ewing, Patrick M. *A Study of the Effects of Individualizing the Pacing and Instruction of Elementary Algebra at the College Level.* (Doctoral dissertation, the Ohio State University, 1973.) Ann Arbor, Mich.: University Microfilms (no. 74-14,509), 1974.

Fernandez, Particia P. *A Presentation and Evaluation of an Individualized Instruction Course in First Year Algebra.* (Doctoral dissertation, University of Utah, 1972.) Ann Arbor, Mich.: University Microfilms (no. 72-24,572), 1972.

Ferney, Gary A. *An Evaluation of a Program for Learning in Accordance with Needs.* (Doctoral dissertation, Washington State University, 1969.) Ann Arbor, Mich.: University Microfilms (no. 70-5657), 1970.

Fielder, Robert E. *The Comparative Effect of Two Years of Individually Prescribed Instruction on Student Achievement in Mathematics.* (Doctoral dissertation, East Texas State University, 1971.) Ann Arbor, Mich.: University Microfilms (no. 72-10,831), 1972.

Finch, John M. *Teaching-Learning Units in PLAN: An Analysis of the Utilization of Instructional Materials to Individualize Learning by Computer-managed Instruction.* (Doctoral dissertation, University of Iowa, 1972.) Ann Arbor, Mich.: University Microfilms (no. 72-26,675), 1972.

Fisher, Jack R. *An Investigation of Three Approaches to the Teaching of Mathematics in the Elementary School.* (Doctoral dissertation, University of Pittsburgh, 1967.) Ann Arbor, Mich.: University Microfilms (no. 68-7841), 1968.

Fisher, Merrill E. *A Comparative Study of Achievement in the Concepts of Fundamentals of Geometry Taught by Computer Managed Individualized Behavioral Objective Instructional Units versus Lecture-Demonstration Methods of Instruction.* (Doctoral dissertation, George Washington University, 1973.) Ann Arbor, Mich.: University Microfilms (no. 73-25,330), 1973.

Fisher, Victor L., Jr. "The Relative Merits of Selected Aspects of Individualized Instruction in an Elementary School Mathematics Program." (Doctoral dissertation, Indiana University.) *Dissertation Abstracts* 27A (1967): 3366.

Flournoy, Lovelia P. "Individualized Instruction in Mathematics for First Grade Children." (Doctoral dissertation, University of California, Los Angeles.) *Dissertation Abstracts* 34A (1974): 5582.

Frase, Larry E. *A Comparison of Two Individualized Mathematics Programs on Student Independence, Achievement, Time, and Attitude Criterion Measures.* (Doctoral dissertation, Arizona State University, 1971.) Ann Arbor, Mich.: University Microfilms (no. 71-26,592), 1971.

Gaskill, Edgar A. *An Evaluation of Individually Prescribed Instruction in the Primary Grades of the Urbana Schools.* (Doctoral dissertation, Illinois State University, 1970.) Ann Arbor, Mich.: University Microfilms (no. 71-6003), 1971.

Gibish, Patricia A. *A Description and Evaluation of the Second Year Implementation of a Systems Approach to Improving Mathematics Instruction.* (Doctoral dissertation, University of Pittsburgh, 1970.) Ann Arbor, Mich.: University Microfilms (no. 71-7995), 1971.

Gilbert, Robert K. *A Comparison of Three Instructional Approaches Using Manipulative Devices in Third Grade Mathematics.* (Doctoral dissertation, University of Minnesota, 1974.) Ann Arbor, Mich.: University Microfilms (no. 75-2099), 1975.

Godde, John A. *A Comparison of Young Children in Achievement of General Skills, Adjustment, and Attitudes, in an Individual Progression Curriculum Organization with Young Children in a Traditional Curriculum Organization.* (Doctoral dissertation, Northern Illinois University, 1972.) Ann Arbor, Mich.: University Microfilms (no. 73-27,589), 1973.

Graham, William A. "Individualized Teaching of Fifth- and Sixth-Grade Arithmetic." *Arithmetic Teacher* 11 (April 1964): 233–34.

Grant, Jettye F. "A Longitudinal Program of Individualized Instruction in Grades 4, 5 and 6." (Doctoral dissertation, University of California, Berkeley.) *Dissertation Abstracts* 25 (1964): 2882.

Hamby, Kelly D. *A Model for Modifying Individualized Instruction.* (Doctoral dissertation, The University of Texas at Austin, 1971.) Ann Arbor, Mich.: University Microfilms (no. 72-19,597), 1972.

Hanneman, James H. *An Experimental Comparison of Independent Study and Conventional Group Instruction in Tenth Grade Geometry.* (Doctoral dissertation, University of Florida, 1971.) Ann Arbor, Mich.: University Microfilms (no. 72-15,689), 1972.

Head, James C. *A Study of the Effectiveness of an Individualized Instruction-Contract Grading Program in a College Algebra Class.* (Doctoral dissertation, George Peabody College for Teachers, 1974.) Ann Arbor, Mich.: University Microfilms (no. 75-12,445), 1975.

Heiman, Marcia B. *Individualized Instruction in the Classroom.* (Doctoral dissertation, University of Michigan, 1970.) Ann Arbor, Mich.: University Microfilms (no. 71-4638), 1971.

Herceg, John. *A Study of the Coordinator's Role in the Introduction of Formally Presented Objectives and Individualized Learning Rates in Computer Assisted Mathematics.* (Doctoral dissertation, University of Pittsburgh, 1972.) Ann Arbor, Mich.: University Microfilms (no. 73-4133), 1973.

Hirsch, Christian R., Jr. *An Experimental Study Comparing the Effects of Guided Discovery and Individualized Instruction on Initial Learning, Transfer, and Retention of Mathematical Concepts and Generalizations.* (Doctoral dissertation, University of Iowa, 1972.) Ann Arbor, Mich.: University Microfilms (no. 73-640), 1973.

Jackson, Grace G. "Continuous Progress in a Primary Unit." (Doctoral dissertation, East Texas State University.) *Dissertation Abstracts* 28A (1967): 2138.

Johnson, Lary. *Minneapolis I.P.I. Mathematics Project 1971–1972; Third Year Evaluation.* Final report of an ESEA Title I project, 1972. (ERIC no. ED 083 290)

Jones, William L. "Comparison of Cognitive and Affective Change of Ninth Grade Students in Open-Space and Closed-Space Classes." (Doctoral dissertation, Arizona State University.) *Dissertation Abstracts* 35A (1974): 1961.

Joyner, Robert N. "The Effect of an NSF-CCSS Project on Junior High School Student Mathematical Achievement and Attitude toward Mathematics." (Doctoral dissertation, Florida State University.) *Dissertation Abstracts* 34A (1974): 5780.

Kontogianes, John T. "The Effects on Achievement, Retention, and Attitude of an Individualized Instructional Program in Mathematics for Prospective Elementary School Teachers." (Doctoral dissertation, University of Oklahoma.) *Dissertation Abstracts* 34A (1974): 5802.

Larsson, Inger. "Individualized Mathematics Teaching Results from the IMU Project in Sweden." *Studia Psychologica et Pedagogica Series Altera* 21 (1973): Also available from ERIC, no. ED 082 997

Lindvall, C. M., and Judy A. Light. *The Use of Manipulative Lessons in Primary Grade Arithmetic in a Program for Individualized Instruction.* Paper presented at the AERA annual meeting, 1974. (ERIC no. ED 090 038)

Lipson, Joseph I. "IPI Math—an Example of What's Right and Wrong with Individualized Modular Programs." *Learning* 2 (1974): 60–61.

Lober, Irene M. "Individually Guided Education-Resource Model." (Doctoral dissertation, Virginia Polytechnic Institute and State University.) *Dissertation Abstracts* 35A (1974): 2589–90.

Ludeman, Clinton, and others. *Final Evaluation Report, Project Video-tape Packages Mathematics.* Final report of an ESEA Title III project, 1973. (ERIC no. ED 086 545)

Malcolm, Paul J. *Analysis of Attitude, Achievement, and Student Profiles as a Result of Individualized Instruction in Mathematics.* (Doctoral dissertation, University of Nebraska, 1972.) Ann Arbor, Mich.: University Microfilms (no. 73-121), 1973.

Matthews, Frank F. *An Investigation of the Feasibility of the Use of Student's Perceived Needs to Control the Rate of Instruction.* Paper presented at the AERA annual meeting, 1974. (ERIC no. ED 091 227)

Mayfield, Irene R. "A Comparative Study: Two Methods of Teaching Mathematics—Conventional and Individualized." (Doctoral dissertation, Mississippi State University.) *Dissertation Abstracts* 34B (1974): 3922–23 .

Morman, Shelba J. "An Audio-Tutorial Method of Instruction vs. the Traditional Lecture-Discussion Method." *Two-Year College Mathematics Journal* 4 (1973): 57–61.

Nanney, Donald L. *The Effects of Individualization and Traditional Mathematics Instruction Programs on Achievement and Self-Concept Scores.* (Doctoral dissertation, University of Miami, 1973.) Ann Arbor, Mich.: University Microfilms (no. 74-14,332), 1974.

Neufeld, K. Allen. *Differences in Personality Characteristics between Groups Having High and Low Mathematical Achievement Gain under Individualized Instruction.* (Doctoral dissertation, University of Wisconsin, 1967.) Ann Arbor, Mich.: University Microfilms (no. 67-16,986), 1967.

Newman, Frederick L., Dennis L. Young, Stanley E. Ball, Clarence C. Smith, and Ronald B. Purtle. "Initial Attitude Differences among Successful, Procrastinating, and 'Withdrawn-from-Course' Students in a Personalized System of Statistics Instruction." *Journal for Research in Mathematics Education* 5 (March 1974): 105–13.

O'Neill, Jane A. *An Analysis on Selected Variables of the Effect of a Systems Approach to Teaching Specific Mathematical Skills to Fifth Grade Students from a Disadvantaged Area.* (Doctoral dissertation, University of Connecticut, 1970.) Ann Arbor, Mich.: University Microfilms (no. 71-16,020), 1971.

Palow, William P. *Modularization—A Road to Relevance?* Paper presented at the meeting of the Florida Junior College Council of Teachers of Mathematics, 1973. (ERIC no. ED 091 232)

Penner, Herbert D. *An Analysis of Using an Individual Progress Approach to the Teaching of Trigonometry in the Omaha, Nebraska, Public High Schools.* (Doctoral dissertation, University of Nebraska, 1972.) Ann Arbor, Mich.: University Microfilms (no. 72-27,419), 1972.

Phillips, Charles A. "A Study of the Effect of Computer Assistance upon the Achievement of Students Utilizing a Curriculum Based on Individually Prescribed Instruction Mathematics." (Doctoral dissertation, Temple University.) *Dissertation Abstracts* 34A (1974): 3710.

Pigford, Valma D. *A Comparison of an Individual Laboratory Method with a Group Teacher-Demonstration Method in Teaching Measurement and Estimation in Metric Units to Preservice Elementary Teachers.* (Doctoral dissertation, Florida State University, 1974.) Ann Arbor, Mich.: University Microfilms (no. 75-941), 1975.

Pond, Thomas F., Jr. *Individualized Instruction: A Model for Teacher Preparation.* (Doctoral dissertation, University of North Dakota, 1973.) Ann Arbor, Mich.: University Microfilms (no. 73-29,630), 1973.

Project Skill—Skill Development through Individual Learning Levels. Final report of an ESEA Title III project, 1972. (ERIC no. ED 082 973)

Pusey, Judith K. "A Comparison of the Effects of Three Instructional Procedures on Achievement, Self-Esteem, and Classroom Adjustment of Intermediate Grade Students in Title I Schools." (Doctoral dissertation, Oklahoma State University.) *Dissertation Abstracts* 34A (1974): 6369.

Putbrese, Larry M. *An Investigation into the Effect of Selected Patterns of Grouping upon Arithmetic Achievement.* (Doctoral dissertation, University of South Dakota, 1971.) Ann Arbor, Mich.: University Microfilms (no. 72-8388), 1972.

Schaefer, William A. *The Relationship of Teaching Methods to Self-Esteem and Achievement in Mathematics among Seventh and Eighth Grade Students.* (Doctoral dissertation, Northern Illinois University, 1972.) Ann Arbor, Mich.: University Microfilms (no. 72-22,803), 1972.

Sherry, Margaret A. *Individualized Contrasted with Traditional Instruction in Sixth Grade Arithmetic Classes.* (Doctoral dissertation, University of Southern California, 1974.) Ann Arbor, Mich.: University Microfilms (no. 75-6445), 1975.

Shumaker, James E. *A Comparison of Study Habits, Study Attitudes, and Academic Achievement in Mathematics in Junior High School of Students Taught by Individually Prescribed Instruction and Students Taught by Traditional Methods of Instruction in Elementary School.* (Doctoral dissertation, University of Pittsburgh, 1972.) Ann Arbor, Mich.: University Microfilms (no. 73-13,176), 1973.

Smith, Gerald E. *The Relationship of Cognitive Style and Instructional Treatment in the Acquisition of Strategies for Teaching Elementary Mathematics.* (Doctoral dissertation, The University of Texas at Austin, 1972.) Ann Arbor, Mich.: University Microfilms (no. 73-520), 1973.

Smith, Jimmy E. "The Effect on Achievement and Attitude of Three Approaches for Developing Area Concepts." (Doctoral dissertation, The University of Texas at Austin.) *Dissertation Abstracts* 34A (1974): 5497–98.

Stone, James L., Jr. *The Effect of Individualized Learning Activity Packages in Mathematics on the Academic Achievement of Seventh- and Eighth-Grade Students in the Demopolis City Schools.* (Doctoral dissertation, University of Alabama, 1974.) Ann Arbor, Mich.: University Microfilms (no. 75-18,305), 1975.

Suydam, Marilyn N., and J. Fred Weaver. *Individualizing Instruction, Set A, Using Research: A Key to Elementary School Mathematics.* Pennsylvania State University Center for Cooperative Research with Schools, 1970. (ERIC no. ED 038 318)

Tack, Robert S. *The Effectiveness of the Westinghouse Learning Center Program Involving a Performance Contract on Reading and Mathematics Achievement of Educationally Deprived Children.* (Doctoral dissertation, Brigham Young University, 1971.) Ann Arbor, Mich.: University Microfilms (no. 72-5766), 1972.

Taylor, Loretta M. *Independent Study versus Presentation by Lecture and Discussion: A Comparative Study of Attitude and Achievement in Two Algebra I Classes.* Doctoral dissertation, University of Northern Colorado, 1971.) Ann Arbor, Mich.: University Microfilms (no. 72-3307), 1972.

Thomas, Bonnie B. *An Evaluation of Individually Prescribed Instruction (IPI) Mathematics in Grades Five and Six of the Urbana Schools.* (Doctoral dissertation, Illinois State University, 1972.) Ann Arbor, Mich.: University Microfilms (no. 72-25,708), 1972.

Thomas, Buren G. *Continuous Progress Advanced Algebra in the Lincoln Public Schools—a Study of Achievement and Attitude toward Mathematics.* (Doctoral dissertation, University of Nebraska, 1971.) Ann Arbor, Mich.: University Microfilms (no. 72-16,019), 1972.

Tychsen, Alfred B. *An Experimental Comparison of Teacher-paced Instruction and Student-paced Instruction in the Teaching of Mathematics in the Public Elementary Schools in Greenwich, Connecticut.* (Doctoral dissertation, University of Connecticut, 1971.) Ann Arbor, Mich.: University Microfilms (no. 71-29,921), 1971.

Verheul, Gustav W. *A Comparison of the Effects of Individually Prescribed Instruction and Conventional Textbook Instruction on Mathematics Learning of Selected Sixth Grade Students.* (Doctoral dissertation, Florida State University, 1971.) Ann Arbor, Mich.: University Microfilms (no. 72-10,052), 1972.

Walters, Ada J. *A Comparison of Pupil Achievement and College Success in Two High School Programs: One Modular and One Traditional.* (Doctoral dissertation, Duke University, 1972.) Ann Arbor, Mich.: University Microfilms (no. 73-6604), 1973.

Wang, Margaret C., and C. M. Lindvall. *An Exploratory Investigation of the Carroll Learning Model and the Bloom Strategy for Mastery Learning.* Pittsburgh, Pa.: Learning Research and Development Center, 1970. (ERIC no. ED 054 983)

Wasden, Frances D. *A Comparative Analysis of the Difference in Achievement between Students Educated in Traditional and Individualized Schools.* (Doctoral dissertation, Brigham Young University, 1971.) Ann Arbor, Mich.: University Microfilms (no. 71-24,270), 1971.

Waters, George H. *The Effects of an Individualized Laboratory Approach on the Teaching of Mathematics to Third Grade Students Achieving below Grade Level.* (Doctoral dissertation, Virginia Polytechnic Institute and State University, 1975.) Ann Arbor, Mich.: University Microfilms (no. 75-11,952), 1975.

Wheaton, Wilbur D. *An Evaluation of an Individualized Learning Program in a California Union High School District.* (Doctoral dissertation, University of Southern California, 1971.) Ann Arbor, Mich.: University Microfilms (no. 72-11,966), 1972.

Whipple, Robert M. *A Statistical Comparison of the Effectiveness of Teaching Metric Geometry by the Laboratory and Individualized Instruction Approaches.* (Doctoral dissertation, Northwestern University, 1972.) Ann Arbor, Mich. University Microfilms (no. 72-32,611), 1972.

Williams, Benjamin G. "An Evaluation of a Continuous Progress Plan in Reading and Mathematics on the Achievement and Attitude of Fourth, Fifth, and Sixth Grade Pupils." (Doctoral dissertation, Lehigh University.) *Dissertation Abstracts* 34A (1974): 7115–16.

Wolff, Bernard R. *An Analysis and Comparison of Individualized Instructional Practices in Arithmetic in Graded and Nongraded Elementary Classrooms in Selected Oregon School Districts.* (Doctoral dissertation, University of Oregon, 1968.) Ann Arbor, Mich.: University Microfilms (no. 69-06,673), 1969.

Wood, Stillman W. *A Study of Selected Student, Instructional, and Achievement Variables within a Program of Individually Prescribed Instruction in Mathematics for Junior High Educable Retardates.* (Doctoral dissertation, University of Oregon, 1972.) Ann Arbor, Mich.: University Microfilms (no. 73-13,782), 1973.

Wright, Robert J. "The Affective and Cognitive Consequences of an Open Education Elementary School." *American Educational Research Journal* 12 (1975): 449–68.

Yomtoob, Youssef. *A Study of the Effect of an Individualized Instructional Program on Attitude, Self-Concepts and Arithmetic Achievement.* (Doctoral dissertation, University of Toledo, 1974.) Ann Arbor, Mich.: University Microfilms (no. 74-29,750), 1974.

13

Hand-held Calculators:
Past, Present, and Future

Max Bell
Edward Esty
Joseph N. Payne
Marilyn N. Suydam

It should come as no surprise that the nonthematic essay in this yearbook is devoted to hand-held calculators. Surely no other device, whether designed as an aid to learning or not, has had more potential for influencing instruction in mathematics. No other topic is more timely for mathematics teachers at all levels. The position statement of the National Council of Teachers of Mathematics, adopted in September 1974, reflects this timeliness:

> With the decrease in cost of the minicalculator, its accessibility to students at all levels is increasing rapidly. Mathematics teachers should recognize the potential contribution of this calculator as a valuable instructional aid. In the classroom, the minicalculator should be used in imaginative ways to reinforce learning and to motivate the learner as he becomes proficient in mathematics.

As we think of the relation between calculators and precollege mathematics, it may be helpful to consider two futures: the immediate future (the one that is inexorably merging into the present) and the long-range future (which remains, at least for a while, comfortably distant).

Of course, there is no firm demarcation between the two: teachers, parents, and administrators are concerned with both. Some of the more pressing questions concern the immediate future: What can I do with calculators in my classroom tomorrow? Should I let my child use the family calculator for doing homework? Should I purchase a classroom set of calculators for my primary team, and if so, what kind should I get? But the long-range questions are there too: How will my fifth-grade year-long calculator program fit into the total mathematics education of my students? As a result of years of calculator usage, will my child be better prepared to deal with real-life mathematical problems or not? Rather than purchase a classroom set of calculators for next fall, should I wait a few years, hoping that prices will continue to drop while mathematical capabilities increase?

It is not the intent of this essay to answer questions like these, for there are no quick answers to give. Indeed, this sampling only hints at the complexity and depth of the issues involved. Instead, our purpose is simply to provide an overview of some recent reports that have dealt in whole or in part with calculators and to list some sources of information that might be helpful to the reader. To those ends, the next four sections are concerned with the report of the National Advisory Committee on Mathematical Education (NACOME), the Euclid Conference report, a report of a status study commissioned by the National Science Foundation, and the report from a calculator conference sponsored jointly by the National Institute of Education and the National Science Foundation. The fifth section is devoted to a potpourri of other calculator activities, and the last section summarizes some of the consistencies across sets of recommendations.

The NACOME Report

The National Advisory Committee on Mathematical Education was appointed by the Conference Board of the Mathematical Sciences in May 1974 to prepare an overview and analysis of school-level mathematical education in the United States—its objectives, current practices, and attainments. Funded by the National Science Foundation, the report of the the committee was released in November 1975.[1]

The topic of hand-held calculators appeared in four separate places in the report. In chapter 2 the strong trend to emphasize computation in the curriculum was severely questioned. The case for decreasing the emphasis on manipulative skills was seen as stronger than ever because of the impending universal availability of calculating equipment. Instead, emphasis was recommended for approximation, order of magnitude, and the interpretation of numerical data. For children who have not acquired functional levels of arithmetic computation by the end of grade 8, the committee suggested that they be provided with calculators to meet their arithmetic needs.

In a separate section on computers, the widespread availability of hand-held calculators was viewed as a challenge to traditional instructional priorities. The following changes were envisioned by the committee (pp. 41–42):

1. The elementary school curriculum will be restructured to include much earlier introduction and greater emphasis on decimal fractions, with corresponding delay and de-emphasis of common fraction notation and algorithms.

2. While students will quickly discover decimals as they experiment with calculators, they will also encounter concepts and operations involving negative integers, exponents, square roots, scientific notation and large numbers—all commonly topics of junior high school instruction. These ideas will then be unavoidable topics of elementary school instruction.

3. Arithmetic proficiency has commonly been assumed as an unavoidable prerequisite to conceptual study and application of mathematical ideas. This practice has condemned many low achieving students to a succession of general mathematics courses that begin with and seldom progress beyond drill in arithmetic skills. Providing these students with calculators has the potential to open a rich new supply of important mathematical ideas for these students—including probability, statistics, functions, graphs, and coordinate geometry—at the same time breaking down self-defeating negative attitudes acquired through years of arithmetic failure.

4. For all students, availability of a calculator does not remove the necessity of analyzing problem situations to determine appropriate calculations and to interpret correctly the numerical results. The user must still determine which calculator buttons to push. With de-emphasis on the purely mechanical aspects of arithmetic comes an opportunity to pay close attention to other crucial aspects of the problem solving process and to treat more genuine problems with the "messy" calculations they inevitably involve. Facility in the mental estimation of arithmetic results, to check that one's calculator is functioning well and that correct problem analysis has preceded calculation, will continue to be useful.

5. Present standards of mathematical achievement will most certainly be invalidated in "calculator classes."

The committee also listed some questions it considered important to investigate through research (pp. 42–43):

When and how should calculator use be introduced so that it does not block needed student understanding and skill in arithmetic operations and algorithms?

Will ready access to calculators facilitate or discourage student memory of basic facts?

For which mathematical procedures is practice with step-by-step paper and pencil calculation essential to thorough understanding and retention?

What types of calculator design—machine logic and display—are optimal for various school uses?

How does calculator availability affect instructional emphasis, curriculum organization, and student learning styles in higher level secondary mathematics subjects like algebra, geometry, trigonometry, and calculus?

In closing the section, the committee expressed the view that calculators would allow students to feel the power of mathematics and would free time for concentration on the conceptual aspects of the subject.

In the final chapter, the following major recommendation was made (p. 138):

That beginning no later than the end of the eighth grade, a calculator should be available for each mathematics student during each mathematics class. Each student should be permitted to use the calculator during all of his or her mathematical work including tests.

The report also identifies particular areas where "new curricular organizations, instructional materials, and courses are of urgent concern," recommends that "instructional materials at all levels" be developed for calculators and suggests that "curricular revision or reorganization in the light of the increasing significance of computers and calculators" is needed (p. 145).

The Euclid Conference

Within the past two years, the Basic Skills Group of the National Institute of Education (NIE) has begun to devote increasing attention to mathematics. It was natural that some of the first questions to arise should concern the nature of basic mathematical skills.

As a first step in sorting out some of the issues, NIE sponsored a Conference on Basic Mathematical Skills and Learning during October 1975 in Euclid, Ohio. Each of the thirty-three participants was asked to submit a position paper before the conference in response to the following questions:

1. What *are* basic mathematical skills and learning?
2. What are the major problems related to children's acquisition of basic mathematical skills and learning, and what role should the National Institute of Education play in addressing these problems?

The final report from the conference consists of two volumes—one containing the thirty-three position papers and the other containing a description of the background and organization of the conference, four working-group reports, and an essay by the conference cochairmen.[2] Not surprisingly, issues related to hand-held calculators were discussed in many of the position papers, in three of the four working groups, and in the essay.[3]

Calculators in the position papers

Although some of the authors mentioned calculators only in a single sentence, others devoted more extensive sections to the subject. The following partial list of topics is presented to indicate the kinds of issues that were treated:

1. The effect of calculators on the curriculum
2. Calculators used as an aid to early counting skills, deficiency in which may later lead to difficulties in addition and multiplication
3. Calculators as a vehicle for a reexamination of the placement and emphasis of various topics in the mathematics curriculum
4. Calculators as an aid to the teaching of programming in grades 5–8
5. The amount and kinds of paper-and-pencil algorithms now needed, both in and out of school
6. The importance of looking toward a future in which hand-held devices will be far more powerful and sophisticated than they are now

Calculators in the working-group reports

Three working groups discussed calculators. The first proposal from the group working on curriculum development and implementation dealt with the impact of calculators and computers. In particular, they recommended studies of—

1. alternative sequences for elementary instruction in arithmetic;
2. uses of the calculator as an aid and stimulus for arithmetic instruction;
3. the impact of calculator availability on problem-solving instruction;
4. the relative importance of various familiar fraction concepts in an environment of calculators (to include an investigation of curriculum topics in later courses such as algebra). [p. 11]

Further, they noted a need at the secondary level to "re-examine curriculum structures and priorities in light of increasing computer and calculator capabilities to perform traditional computations," saying that "this clearly affects the definition of basic skills" (p. 12).

The group working on goals for basic mathematical skills and learning noted that "the whole issue of the effect of the calculator on the teaching of arithmetic is a very complex one which deserves considerable investigation and consideration" (p. 18). The problem is one of finding the proper balance between a minimum of single-digit arithmetic and the amount of paper-and-pencil computation that is presently taught. The group concluded that it is a "question that needs further study and far more discussion among a broader base of people" (p. 18).[4]

The third working group to mention calculators was concerned with research priorities. It recommended as one of the "first priority questions of practical urgency" the following (p. 29):

> *True functional needs (remember Xerox).* In the midst of controversy about what skills people ought to have, this question calls for the matter to be studied empirically—track people's behavior and *see* where they need which skills. "Remember Xerox" reminds us that "needs" should not merely mean needs exercised by current practices. As with the office copier, there may be unrecognized needs. In particular, the potential of hand calculators must be explored. What are their innovative uses? Learning uses? What would be the long-range implications of substituting calculators for pencil-and-paper computation?

Calculators in the essay

Finally, one of the ten sections of the essay by Hilton and Rising was devoted to calculators. They cautioned against placing any premature restraints on calculator usage—restraints either on grade level or on machine logic—fearing that interesting avenues of research would otherwise be blocked. They noted a need for good curriculum materials, particularly to "support and extend conceptual understanding of mathematics and to facilitate the application of arithmetical techniques to the solution of real life problems" (p. 39).

The Report to NSF

In March 1975, the National Science Foundation, concerned about the potential impact of the calculator on the precollege mathematics curriculum, funded an investigation involving a critical analysis of the role of the calculator.[5] The study was designed to identify the range of beliefs and reactions about calculators and in particular the arguments that were being used to support positions strongly favorable and strongly negative toward the use of calculators in elementary and secondary schools.

Questionnaires were sent to teachers and other school personnel, to state supervisors of mathematics, to mathematics educators in colleges and universities, and to textbook publishers. In addition to asking for statements of arguments for and against the use of calculators, the form contained such questions as these:

How should calculators be used?

What uses are most important at various levels?

How should the curriculum be modified if calculators are readily available to students at all times?

What would you recommend to elementary and secondary school personnel considering the selection and use of calculators?

Appendix B of the report contains the complete sets of responses from questionnaires.

Articles in educational and noneducational journals and in newspapers, reports from calculator manufacturers, curriculum materials, position papers, conference reports, and other documents were surveyed for additional arguments. (Appendix A of the report contains an annotated list of these references.) Manufacturers were surveyed in an attempt to secure information on current and future sales and development. Research reports were checked to determine what had already been ascertained about the effect of using calculators.

In addition to the analysis in the final report, several position papers were prepared by persons who have devoted much thought to the promises and problems posed by the use of calculators in elementary and secondary schools. The position papers are incorporated as appendixes C, D, E, and F: appendix C, *Teaching Mathematics with the Hand-Held Calculator,* by George Immerzeel, Earl Ockenga, and John Tarr; appendix D, *Hand-Held Calculators and Potential Redesign of the School Mathematics Curriculum,* by H. O. Pollak; appendix E, *Some Suggestions for Needed Research on the Role of the Hand-Held Electronic Calculator in Relation to School Mathematics Curricula,* by J. F. Weaver; and appendix F, *Calculators and School Arithmetic: Some Perspectives,* by Zalman Usiskin and Max Bell.

The case for using calculators in schools

The analysis of positions that people hold regarding the use of calculators in schools made it apparent that there is much similarity in viewpoints. In addition, these same viewpoints are reflected over and over in published articles. The most frequently cited reasons *for* using calculators in schools were these:

1. They aid in computation. They are practical, convenient, and efficient. They remove drudgery and save time on tedious calculation. They are less frustrating, especially for low achievers. They encourage speed and accuracy.

2. They facilitate understanding and concept development.

3. They lessen the need for memorization, especially when used to reinforce basic facts and concepts with immediate feedback. They encourage estimation, approximation, and verification.

4. They help in problem solving. Problems can be more realistic, and the scope of problem solving can be enlarged.

5. They motivate. They encourage curiosity, positive attitudes, and independence.

6. They aid in exploring, understanding, and learning algorithmic processes.

7. They encourage discovery, exploration, and creativity.

8. They exist. They are here to stay in the real world; so we cannot ignore them.

The last reason—the pragmatic fact that calculators exist and that they are appearing in the hands of increasing numbers of students—is perhaps the most compelling. *How* they can be used to facilitate each of the other seven beliefs is therefore a question that must be answered.

The case against using calculators in schools

The most frequently cited reasons for *not* using calculators in schools were these:

1. They could be used as substitutes for developing computational skills: students might not be motivated to master basic facts and algorithms.

2. They are not available to all, and so some students are at a disadvantage.

3. They may give a false impression of what mathematics is. Mathematics may be equated to computation, performed without thinking. Emphasis is on the product rather than on the process; structure is de-emphasized. Mental laziness and too much dependence are encouraged; a lack of understanding is promoted. Some students and teachers will misuse them.

4. They are faddish. There has been little planning or research on their use in classrooms.

5. They lead to maintenance and security problems.

The first concern—that students will not learn basic mathematical skills—was one expressed most frequently by parents and by other members of the public, as reflected by newspaper articles. But few educators believe that children should use calculators in place of learning basic mathematical skills. Rather, they express a strong belief that calculators can help children develop and learn more mathematical skills and ideas than would be possible without the use of calculators. Much serious attention must be given by teachers and others to proving that this belief can be supported and become fact.

Empirical evidence

In his position paper on needed research, prepared for inclusion with the report, Weaver pointed out that—

> the very newness of calculators provides little of a research base upon which to build. . . . The extent of ongoing research is very difficult to assess; this also is true of the nature of that research. We are given hints from the brief progress reports released by some projects . . . but by and large we have precious little information—and none of it definitive— regarding the extent and nature of ongoing research. [p. 18]

Most of the studies to date have been exploratory. Some of the "hardest" data come from studies conducted by calculator manufacturers; not surprisingly, these indicate that students (*a*) can use the calculator with a variety of content and (*b*) achieve well when using the calculator. Many schools are checking data on their own students to determine the effect of the use of calculators: reports indicate that using calculators generally results in achievement at least as high as that which results when calculators are not used; in some instances, computation scores are significantly higher when calculators are used, and in others problem-solving scores are significantly higher.

A summary of suggestions for research that should be conducted to determine the potential and the problems of using calculators in schools includes investigations related to the following:

- When and how to introduce calculators
- Effective procedures for learning basic facts, computational skills, problem solving, and various mathematical ideas
- Effective algorithms for calculators
- Long-range effects of using calculator algorithms
- The need for paper-and-pencil algorithms
- The effect of using calculators with specific content and curricula, including the effect of curricular sequence and changes of emphasis
- The relationship between work with calculators and work with computers
- Changes in teacher-education curricula
- Optimal designs and functions of calculators

One very specific caution must be emphasized: attempts to restructure the curriculum, either extensively or minimally, *must not* proceed independently of research. The two are interwoven, and one cannot be effective without the other.

Recommendations for curriculum and instruction

A variety of recommendations were suggested by educators responding to the questionnaire. These recommendations ranged from the general to those supporting specific curriculum changes. For example:

1. Experiment and plan, finding meaningful ways to use calculators. Develop a school-wide policy and guidelines, incorporating calculators into the existing curriculum and developing new curricula as necessary. Plan a reasonable in-service program, evaluation, and research to support the use of calculators.

2. Survey available calculator models carefully, and buy good equipment commensurate with student needs. Make sure that all students have access to a calculator.

3. Think of calculators as tools to extend mathematical understanding and learning by making traditional work easier. The focus can be on process because the product is assured.

4. Change teaching emphases to concept development, algorithmic processes, applications of various operations, and problem solving using real-life and interdisciplinary applications. Place more emphasis on problem-solving strategies. Use practical, realistic, significant problems and more applications.

5. Do not ignore the development of computational skill. Spend less time on computational drill and more time on concepts and the meaning of operations. Use more laboratory activities where computation is involved but the emphasis is on learning mathematical concepts. Decrease the use of tedious, complicated algorithms; emphasize algorithmic learning, including the development of algorithms by students.

6. De-emphasize fractions and emphasize decimals, introducing them earlier.

7. Emphasize estimation and approximation (including mental computational skills), checking and feedback, exploration and discovery. Do more or earlier work with such ideas as place value, the decimal system, number theory, number patterns, sequences, limits, functions, iteration, statistics, probability, flowcharting, computer literacy, large numbers, negative numbers, scientific notation, data generation, and formula testing.

Position papers

The four position papers in the report present viewpoints indicating the need for both immediate and long-range planning. Immerzeel, Ockenga, and Tarr point out that to avoid "future shock" imaginative software must be developed. They make recommendations and provide a variety of specific illustrations for using the calculator at each of several levels, usually with topics from existing curricula.

Pollak describes two partial orderings that are often used in designing a mathematics curriculum. The first is dependent on the structure or nature of mathematics and may be called *content ordering*. A second partial ordering that must be considered in curriculum development is *societal ordering;* that is, topics in a mathematics curriculum are included and ordered according to the topic's worth to society.

If the introduction of a new device such as the hand-held calculator makes a significant change in either the content ordering or the societal ordering, then major curriculum modifications seem appropriate. For example, the algorithm for addition with decimals is the same as the algorithm for addition with whole numbers: the same buttons are pushed for each. Thus, it may no longer be necessary or desirable to delay the teaching of decimals until fifth grade.

A careful, extensive study of the impact of the calculator on the curriculum is needed: it seems that significant changes could or ought to be made. In their position paper, Usiskin and Bell present some initial suggestions on this task. They take exception to merely incorporating the calculator into existing curricula: "It is thus our belief that the insertion of calculators into K–6 classrooms using most existing curricula is fraught with peril" (p. 36). They argue for an alternative curriculum and provide an assessment of how the curriculum could be restructured. They note that those who view the present curriculum as optimally logical and sequential may find this threatening; their specific suggestions could, however, suggest to many teachers a different way of considering the use of calculators.

Closely related to both curriculum development and instructional concerns is the role of research. Weaver discusses some of the research questions that should have priority. He called attention to two points:

> The greatest thing we have to fear today about the calculator vis-á-vis school mathematics curricula is the degree of fear that already exists about the calculator vis-á-vis school mathematics curricula. [p. 2]
>
> To many persons the calculator threatens to violate certain tenets regarding school mathematics learning and instruction. . . . Sugestions for calculator uses are made within the constraints of those tenets . . . and any research that might be implicit in such suggestions would be similarly constrained. . . . Some other persons, however, appear to be willing—and possibly even anxious—to suggest calculator uses that may challenge certain of our cherished tenets. [p. 5]

Weaver distinguishes among three types of curricula—calculator-assisted, calculator-modulated, and calculator-based—and points out that "research should not be unmindful of such differential roles." He discusses six research questions for which answers should be sought; each in turn can lead to a series of investigations.

Conference on the Uses of Hand-held Calculators in Education

The Conference on the Uses of Hand-held Calculators in Education was convened in Washington, D.C., in June 1976 by the National Institute of Education and the National Science Foundation. Participants were charged to "produce a . . . planning document that will provide a well-defined framework for future research and development efforts . . . that NSF and NIE may use if they wish as a guide to program planning."

In addition to the conference participants, seven persons attended the conference briefly to make presentations. They ranged from representatives of the microelectronics industry with a rather intense "get with it lest the world leave you behind" message to a representative of a citizens group so disturbed about possible negative effects as to have already petitioned a state legislature for outright prohibition of calculators in schools.

The participants concluded early that calculators are certain to be a significant fact of life in the world outside of schools and that they overlap so heavily with what is done in school that schools may not be permitted any *choice* about taking account of them. That is, substantial influence of calculators on schools was seen as inevitable, for better or worse. With that conclusion, the work of the conference turned to the outlining of research and development to increase the probability of better rather than worse consequences.

Conference participants agreed that the role of the conference was not to try to *answer* questions about the effects of calculators in schools but rather to *identify* such questions and suggest initiatives in information gathering, research, and development. They were particularly concerned that a variety of questions be kept open for investigation that might otherwise be foreclosed by too hasty conclusions. They tried to sort out what initiatives in research and development might be fruitful in helping teachers, youngsters, and parents cope with the immediate challenges presented by calculators. But they also tried to identify ideas that might result in substantial new approaches to teaching mathematics.

The conference report summarized discussion about many aspects of present-day school mathematics and the opportunities and dangers presented by calculators. The recommendations that emerged from those discussions are synthesized here. For the specific recommendations, with their rationale and other details, a copy of the report should be obtained.

The recommendations are grouped into several broad areas of concern: dissemination, the development of an information base, the development of curriculum materials for the immediate future and for the long-range future, research and evaluation considerations, and teacher education.

Dissemination

A central information collection and dissemination center should be established to make easily available both samples of materials being produced and research reports related to the use of calculators in schools. The means to better dissemination of information, both to the public and to the teaching profession, should be explored.

Development of an information base

A set of studies should be commissioned to provide intensive and critical analyses of existing mathematics curriculum materials, current instructional practices, and actual uses of computation (and other mathematics). Such information would provide a basis for curriculum development and research efforts involving the calculator. In addition, calculator-relevant information related to different aspects of mathematics learning should be critically reviewed. A continuing study should monitor the calculator as an educational innovation.

Curriculum development for the immediate future

Many curriculum development and research efforts focused on particular topics and small components of courses or curricula should be undertaken both to fill immediately pressing needs in accommodating to calculators and to pioneer new work that may later become part of considerably restructured courses. Attention is directed in particular to the role of algorithms and algorithmic processes, problem solving, a variety of topics such as estimation, and topics not presently taught at particular levels.

Curriculum development for the long-range future

A variety of full-scale alternatives to current school mathematics programs, fully integrating the use of calculators, should be developed. Psychological and other theories or models should not be neglected in the development of these alternatives.

Research and evaluation considerations

Research and evaluation related to calculators must interact strongly with initiatives in curriculum development. Sampling and other aspects of research methodology should be given careful consideration. There should be development and research focused on new approaches to achievement testing and to assessing attitudes and other noncognitive outcomes of school mathematics instruction. Both achievement and attitude in relation to the use of calculators should be carefully assessed, including any change in the amount of time spent in mathematics instruction. Means should be found to assure truly critical evaluations of calculator-oriented materials (and calculators themselves) proposed for school use.

Teacher education

The training and retraining of teachers to help them respond to calculators and calculator-influenced curriculum materials must be made nearly equal in importance to the development of new initiatives aimed at school students.

Other Activities

Many schools are incorporating calculators into existing mathematics programs. Most of them seem to be proceeding with caution because of parental concern that calculators may have an adverse effect on achievement. To help allay these genuine concerns, schools are keeping rather careful records on achievement.

One specific example is from Livonia, Michigan. During the 1976–77 school year, five different fourth-grade classrooms were provided with hand-held calculators, five in each room. Each child had regular access to a calculator for about twenty minutes each week. It could be used in doing classwork or in other ways. Achievement data on a standardized mathematics test from these five classes will be compared with achievement data from five comparable control classes. If the results seem promising, the next question expected to be explored is, How can the calculators be integrated into the mathematics instruction in the classes? The proportion of class time devoted to calculators may be increased also if the data seem to support it.[7]

Also in Livonia, four units designed for use with hand-held calculators have been written in a self-teaching format. They are planned for use in a consumer mathematics course and a remedial mathematics course for students in grades 10, 11, and 12. If the pilot use of the units during the year seems promising, it is likely that five or six additional units will be written.

Another example, among many that could be cited, is from Western Springs, Illinois. There, with the aid of a parent advisory group, School District 101 is conducting a study in which the calculator is used throughout the district as a teaching aid at the fifth- and sixth-grade levels. The district is seeking the answers to five questions involving the effect of the use of the calculator on achievement, attitude, and motivation; how the calculator can best be used; and what curriculum changes are needed. In some fifth-grade rooms, each child has access to his or her own calculator, and highly directed units of study have been developed to use the calculator as the primary mode of instruction on whole numbers, fractions, decimals, and metric measurement. In other fifth-grade rooms, a limited number of calculators are available to use as supportive tools for checking paper-and-pencil computation. Some sixth-grade classes have five or more calculators for use in a nondirected, exploratory program. Several classes at both levels act as control groups. Testing indicated that the group using calculators with calculator-oriented materials scored significantly higher than the other three groups; the group using calculators in a supportive role scored higher than the remaining two groups. Implications for future efforts might be evident as data amasses.[8]

In general, certain patterns concerning the use of calculators are evident:

1. The district or school purchases a small number of calculators, which are given to teachers for exploratory activities. This is followed by discussion and decision on whether the district or school should purchase more (e.g., Macomb County, Mich.; Columbus, Ohio).

2. Remedial mathematics or Title I classes receive calculators for use with low achievers who have not previously learned computational skills well (e.g., Washington Irving High School, New York City; Berkeley, Calif.).

3. Calculators are placed in advanced science and mathematics classes in secondary schools (e.g., Lubbock, Tex.).

4. Pilot studies and research are being conducted on the effect of the use of calculators (e.g., with low achievers at the secondary level in Chicago; in schools in such California cities as Cupertino, Garden Grove, Los Angeles, San Diego, San Francisco, and Santa Barbara).

5. Teachers and students bring calculators into the classroom and use them when it seems feasible (instances too numerous to mention).

Among the variety of other activities, surveys of the attitudes of teachers toward calculators and of the uses being made of calculators have been conducted (e.g., in Philadelphia, Pa.; Shawnee Mission, Kans.; and the states of Ohio and California); these sometimes lead to the development of policy statements. Statewide conferences in, for instance, Michigan and Ohio have focused on the role of calculators; workshops have been presented at local, regional, and national mathematics meetings throughout the country. The National Council of Teachers of Mathematics has assembled in Reston, Virginia, a collection of calculators and materials for use with calculators and has developed a set of three films illustrating ways of using calculators.[9] In cooperation with the ERIC Center for Science,

Mathematics, and Environmental Education, NCTM has also prepared a compilation of teacher-suggested activities for use with calculators.[10] Several journals (e.g., the *Arithmetic Teacher*[11]) have had issues focused on the calculator. Some state professional organizations have produced collections of activities for use at various levels. For instance, the Iowa Council of Teachers of Mathematics has produced *The Hand-held Calculator*,[12] and the Michigan Council of Teachers of Mathematics prepared *Uses of the Calculator in School Mathematics, K–12*.[13] The interest and involvement of an increasing number of teachers, parents, and organizations is evident.

Summary

A variety of activities and recommendations have been reviewed. The intent of this overview is to inform the reader of some sources of information that are available: each cited document includes specific thoughts and ideas for action.

As the recommendations of various groups are analyzed, it is apparent that there is a high degree of overlapping. Across these groups, which include classroom teachers, supervisors, mathematics educators, mathematicians, parents, and others, there seems to be agreement that—

1. the calculator cannot be ignored: its use in the schools must be carefully explored;
2. the focus of attention must be on how the calculator can *best* be used to develop and reinforce mathematical skills and ideas;
3. studies must be made of the uses of mathematics needed by children and by adults, given widespread use of the calculator in our society;
4. research and curriculum development must proceed hand in hand;
5. although some attention must be given to immediate concerns, this must not preclude long-range planning for mathematics curricula and instructional practices that incorporate optimal use of the calculator.

In the task of evolving new materials, practices, and curricula, everyone who teaches and uses mathematics has a role to play. We are all learners in this process of adapting to change.

NOTES

1. Copies of the NACOME report, entitled *Overview and Analysis of School Mathematics: Grades K–12,* are available through the ERIC Document Reproduction

Service (EDRS), P.O. Box 190, Arlington, VA 22210. The October 1976 issue of the *Mathematics Teacher* contained a summary article as well as several individual reactions to the NACOME report.

2. Single copies of the Euclid Conference report are available free while the supply lasts from E. Esty, Mail Stop 7, NIE, 1200 Nineteenth St., NW, Washington, DC 20208. The report can also be obtained through EDRS.

3. For the convenience of those who have access to a copy of the Euclid Conference report, references to hand-held calculators are found on the following pages: volume 1: 5, 24–26, 45, 60, 76, 94, 101–4, 119, 127, 151–52, 159, 173, 175, 203–4, 213, 223–24; volume 2: 11–12, 17–18, 29, 38–39.

4. One professional group that has been studying these questions is the National Council of Supervisors of Mathematics. NCSM conducted its own conference in July 1976 as a follow-up to the Euclid Conference; its report will be available in April 1977.

5. The NSF report, by Marilyn N. Suydam, is *Electronic Hand Calculators: The Implications for Pre-College Education,* NSF Grant No. EPP 75-16757, February 1976. Copies of the report are available from EDRS.

6. Information on the report of the NIE/NSF conference can be obtained from E. Esty, Mail Stop 7, NIE, 1200 Nineteenth St., NW, Washington, DC 20208. The report can also be obtained through EDRS.

7. Information on the results can be obtained from Charles J. Zoet, Mathematics Coordinator, Livonia Schools, 15125 Farmington Rd., Livonia, MI 48154.

8. Further information can be obtained from Kay Nebel, Project Director, School District 101, 4225 Wolf Rd., Western Springs, IL 60558.

9. Further information on the films, including rental fees or purchase price, is available from the distributor, Encyclopedia Britannica Education Corporation, 425 N. Michigan Ave., Chicago, IL 60611.

10. The publication is available from the National Council of Teachers of Mathematics, 1906 Association Dr., Reston, VA 22091.

11. The November 1976 issue of the *Arithmetic Teacher* contained articles on a variety of calculator applications.

12. The publication is available for $2 from Ann Robinson, ICTM, 2712 Cedar Heights Dr., Cedar Falls, IA 50613.

13. The publication is available for $2 from Horace L. Mowrer, MCTM Publications Chairman, 2165 E. Maple Rd., Birmingham, MI 48008.